Data Analysis in the
Earth Sciences
Using
MATLAB®

Gerard V. Middleton

McMaster University

Prentice Hall
Upper Saddle River, New Jersey 07458

Library of Congress Cataloging-in-Publication Data

Middleton, Gerard V.
Data analysis in the earth sciences using MATLAB® / Gerard V. Middleton.
　　p. cm.
　Includes index.
　ISBN 0–13–393505–1
　1. Earth sciences–Statistical methods–Data processing.
　2. MATLAB.　I. Title.
QE33.2.S82M53　2000
550'.72—dc21　　　　　　　　　　　　　　　　　　　　99–21690
　　　　　　　　　　　　　　　　　　　　　　　　　　　　CIP

Senior Editor: *Patrick Lynck*
Executive Managing Editor: *Kathleen Schiaparelli*
Assistant Managing Editor: *Lisa Kinne*
Executive Marketing Manager: *Leslie Cavaliere*
Production Editor: *Kim Dellas*
Manufacturing Manager: *Trudy Pisciotti*
Art Director: *Jayne Conte*
Cover Designer: *Bruce Kenselaar*
Editorial Assistant: *Betsy Williams*

MATLAB® is a registered
trademark of The MathWorks, Inc.

Printed in the United States of America

10　9　8　7　6　5　4　3　2　1

ISBN　0-13-393505-1

Prentice-Hall International (UK) Limited, *London*
Prentice-Hall of Australia Pty. Limited, *Sydney*
Prentice-Hall Canada Inc., *Toronto*
Prentice-Hall Hispanoamericana, S.A., *Mexico City*
Prentice-Hall of India Private Limited, *New Delhi*
Prentice-Hall of Japan, Inc., *Tokyo*
Prentice-Hall (Singapore) Pte. Ltd., *Singapore*
Editora Prentice-Hall do Brasil, Ltda., *Rio de Janeiro*

Contents

Preface

Earth scientists deal with a variety of different types of numerical data, generated as measurements made in the laboratory or in the field. Some data-generating projects are on a very large scale, and require correspondingly complex, and often custom-designed, computer programs. Most, however, are of limited scope, and require general purpose data reduction and graphical display, plus a few more-or-less standard statistical manipulations. Earth scientists commonly use spreadsheets for this type of work—but spreadsheets are not designed for scientific applications, and programming them to perform such operations as contouring, calculating Fourier transforms, or plotting triangular diagrams, rose diagrams, and stereographic nets is not at all straightforward. This book demonstrates the use in the earth sciences of a commercial software package called MATLAB®. MATLAB is a registered trademark of The MathWorks, Inc.

This book is addressed to two groups of readers:

- Students need a book which illustrates the potential uses of MATLAB. Students' background in statistics and computing tends to be highly variable: some may have already taken an introductory course in statistics; most have used a personal computer for such tasks as word processing, but few have any extensive experience in programming. Their course in data analysis should attempt to broaden their computing experience, at the same time as it introduces them to some of the most useful numerical techniques—few of which are taught in introductory statistics courses. There are several good texts (e.g., Davis, 1986; Carr, 1995; Swan and Sandilands, 1995) which provide a more extensive discussion of the mathematical techniques than we attempt in this book. Only one of these (Carr) gives illustrative computer source code.

- Professional earth scientists need a book which illustrates the potential uses of MATLAB in the earth sciences, but they are likely to be familiar with at least some of the numerical techniques discussed in this book. An extended treatment of elementary descriptive statistics and data processing is not needed for this group of readers.

For students, the brief treatment of numerical techniques given in this book should be supplemented either by the instructor's lectures, or by a reference text, or both. The combination of the Manual provided with the Student Edition of MATLAB, and the examples given in this book, should be adequate to give students a facility with MATLAB that will

vii

serve them well in their professional work. Unlike some software designed for students (e.g., Minitab), the Student Edition of MATLAB is essentially the same powerful program that is available on a wide variety of personal computers and workstations, and produces excellent printed graphics.

For professional earth scientists, I hope to provide some examples to demonstrate that most of their day-to-day needs in numerical scientific computing are better carried out using MATLAB than spreadsheets, and that it will be easier to use MATLAB for this purpose than to try and assemble a large number of special purpose programs (many of which do not supply source code, or supply source code in unfamiliar languages, so that modification is difficult). Of course, advanced work in almost any area touched on in this book will require more, and more sophisticated, functions than I have presented. Some of these are available in the toolboxes available from The MathWorks, Inc. (e.g., the Statistics, Signal Processing, and Spline Toolboxes); others are available as free- or shareware. The examples in the book were originally written in MATLAB, version 4, but have been tested, and modified where necessary for version 5.

MATLAB comes with excellent manuals. The reader of this book is urged to complete the tutorial sections of these manuals before using and (particularly) modifying the scripts listed in this text. Reading the scripts is an integral part of reading this text—they are well annotated with comments, explaining their function, and they should not be skipped (which is why they are not relegated to appendices).

The MathWorks, Inc. maintains an `ftp` site (`ftp.mathworks.com`) and a Web site (`http:\\www.mathworks.com`) from which many useful supplementary functions and scripts may be obtained (see particularly those in the `\pub \contrib` directories). Some are referenced in the text, but the number will change and no doubt grow rapidly as the number of MATLAB users grows in the earth sciences. Almost all the scripts and data files used in this book are also available from this source. I have tried to keep the book short by including only a few of the sets of data as tables: students need to prepare a few short data files as an exercise in using an editor, but they (and other readers) certainly do not need to experience the tedium of typing long data tables into the computer.

Inquiries about MATLAB may be directed to:

> The MathWorks, Inc.
> 24 Prime Park Way
> Natick, MA 01760-1500
> Tel: 508-647-7000
> Fax: 508-647-7001
> EMail: info@mathworks.com
> WWW: www.mathworks.com

I believe that most readers will find it advisable to purchase a mathematical reference work, which will be useful to them throughout their professional lives. Among those that I have found most useful is:

Gellert, W., H. Küstner, M. Hellwich, and H. Kästner, eds., 1975, The VNR Concise Encyclopedia of Mathematics. New York, Van Nostrand Reinhold, 760 p.

This book has been prepared using LaTeX2e, in the PC version made available by the Dutch TeXUser's Group on CD-ROM (4allTeX—for more information Email `ntg@nic.surfnet.nl`). TeX and LaTeX are computer typesetting programs developed by Donald Knuth and Leslie Lamport. The PC implementation (emTeX) was written by Eberhard Mattes. All of these (as well as versions for other computers) are available free of charge, or at nominal cost. LaTeX is now adopted by most mathematics and physics journals, the Royal Society (of London), and by the American Geophysical Union. Its use essentially eliminates the cost of typesetting, as well as providing a common language for the exchange of symbol-rich information between scientists. (The author believes that some introduction to the use of LaTeX should also be a part of the earth science curriculum. LaTeX commands are also used to obtain some symbols in the latest version of MATLAB.) We all owe a debt of gratitude to those who have made such a sophisticated system available without taking any personal profit.

The author thanks Pierre Brassard, Denis Shaw, Eric Breitenberger, Peter Wilcock, Martin Trauth, Jochen Roessler, and the Prentice Hall reviewers Craig Borghesani (Terasoft, Inc.), Jim Carr (Mackay School of Mines, University of Nevada, Reno), and Erich U. Peterson (University of Utah) for their comments on earlier drafts of the manuscript. Denis Shaw also referred me to several papers and sources of data that were previously unknown to me: our discussions on the problems of data processing in the earth sciences go back more than 40 years, so I probably owe him for many more favors than I now remember! Special thanks are due to the students who took the course at McMaster, and convinced the author that a semester course combining data analysis concepts and MATLAB use was a real possibility. Unfortunately, due to cutbacks in funding to Ontario universities, since the author's retirement this course no longer exists at McMaster—but perhaps our experience can be passed on to others who have not been forced to cancel courses to meet downsized budgets.

I would like to dedicate this book to the memory of the pioneers who inspired my first steps in data analysis, some 40 years ago: J.C. Griffiths, W.C. Krumbein, and R.L. Miller.

Gerard V. Middleton
McMaster University

About the Author

Gerard V. Middleton is Professor emeritus of Geology at McMaster, and a co-author of "Origin of Sedimentary Rocks" (Prentice-Hall), "Mechanics in the Earth and Environmental Sciences" (Cambridge), and "Nonlinear Dynamics and Fractals: New Techniques for Sedimentary Data" (SEPM). His computing experience dates back to the early sixties, when the first computer (a Bendix G-15) was delivered to McMaster. In 1998 he received the "Distinguished Educator Award" of the American Association of Petroleum Geologists.

Chapter 1

Introduction

1.1 This Book

This book is about data in the earth sciences: its nature, its graphic representation, and its numerical analysis. The book makes use of MATLAB as the main software for graphics and numerical analysis. Why MATLAB?

MATLAB is a commercial package for numerical analysis, developed by The Math-Works, Inc. It is based on the concept that data can be represented as arrays of numbers, either in rows or columns (one-dimensional arrays are called *vectors*) or both (two-dimensional arrays are called *matrices*). Then most of the operations of numerical analysis can be carried out using linear (or matrix) algebra. Some of these operations are quite complicated to program using ordinary programming languages like BASIC, FORTRAN, Pascal, or C—but they have been incorporated as single commands in MATLAB. MAT-LAB also provides a versatile series of commands for producing and printing high-quality graphical representations of data. This relieves the user of two of the major problems that anyone meets who tries to write even the simplest data-analysis program.

MATLAB has been implemented for many different types of computers, not only the PCs and Macintoshs that a student is likely to encounter, but also the more powerful workstations that may be encountered later as a professional working in the earth sciences. Thus the effort expended learning the PC or Macintosh version is an investment in the future: essentially the same software can be used for professional work. The inexpensive Student Version of MATLAB is very like the professional versions, and is accompanied by a manual containing an excellent tutorial introduction. It also contains parts of some Toolboxes available only as add-ons for the professional versions. It is restricted only by the size of the data arrays that can be handled. The draft for this book was written using Version 4 for PC: the published version has been upgraded to Version 5. Version 4 ran on PCs with WINDOWS® 3.1. Version 5 runs only on PCs with WINDOWS® 95. Not everyone has WINDOWS 95, so as far as possible, the book discusses features that are common to both versions, but it includes comments on features of Version 5 that are of special interest in the earth sciences.

Why a special book about using MATLAB for data analysis in the earth sciences?

Data in the earth sciences differ in some important ways from data in other branches of science. The earth sciences often deal with series of measurements made sequentially in time or space, or with spatially distributed data. Also the data may be directional, for example, velocities which measure not only the magnitude (speed) but also the direction of flow, or measurements of the dip and strike of strata, or the orientation of other rock structures. Special techniques of data analysis have been devised to deal with such data, but they are rarely treated in texts on elementary statistics. Some of the techniques needed (e.g., Fourier analysis, contouring) are already implemented in MATLAB, and others (e.g., rose diagrams, stereographic projections) will be implemented in original MATLAB scripts prepared for this book.

1.2   Data in the Earth Sciences

The earth sciences are a diverse group of disciplines, ranging from those concerned almost exclusively with fluids (meteorology, oceanography) to those concerned mainly with solid minerals or rocks. In each discipline research workers have sought to isolate the most important aspects under study, and devise appropriate techniques of measurement. Fluids are characterized by their chemical composition, and physical properties, such as temperature, density, and velocity; minerals are classified by their atomic structure and chemical composition; soils, sediments, and rocks are described by their mineralogical or chemical composition, structures, and textures.

It is useful to distinguish four scales of measurement (Stevens, 1946):

- **Nominal** Most people would call this "classification" rather than true measurement. For example, grains in a sand can be identified as belonging to one of several categories (quartz, feldspar, rock fragments, etc.); organisms (living or fossil) can be classified by genus or species. In a list, each category might be assigned a number (e.g., quartz = 1, feldspar = 2, etc.) but these numbers are used only for identification: it makes no sense to say that "two quartzes equal one feldspar."

- **Ordinal** This is also a form of classification but now the ordering of the categories is important. For example, sediments in a deep-sea core might be categorized by age as (1) Holocene, (2) Pleistocene, and (3) Miocene. The order indicated by the numbers is significant, but the Pleistocene sediments are not necessarily twice as old as the Holocene, and the difference in age between the Holocene and the Pleistocene is not the same as the difference in age between the Pleistocene and the Miocene. A good example of an ordinal scale, taught in most introductory geology courses, is Mohs' scale of mineral hardness: talc is assigned a hardness of 1, gypsum 2, and so on up to a hardness of 10 for diamond. Minerals higher up the hardness scale can scratch those lower in the scale, but the intervals of hardness are known to be unequal.

- **Interval** This is the first scale of measurement that makes use of most of the properties of the real numbers. A good example is the Celsius scale of temperature. The interval between 0 and 1 on the scale is the same as the interval between 1 and 2 on the scale (it would take us too far into thermodynamics to define exactly how we know this!), but it is still incorrect to say that a body at 2° C is twice as hot as one at 1° C. This is because the Celsius scale has an arbitrarily chosen zero.

- **Ratio** In this scale not only are the intervals between numbers equal, but so are their ratios. In the Kelvin scale of absolute temperatures, it makes thermodynamic sense to say that a body at 400 K is twice as hot as one at 200 K. If samples from a deep-sea core can not only be identified from their fossil content as Holocene, Pleistocene, and Miocene, but also assigned "absolute ages" (from radioactive dating techniques) of 5,000, 100,000, and 2 million years, then it makes sense to say that the Pleistocene sample is 20 times as old as the Holocene sample, and the Miocene sample is 20 times as old as the Pleistocene sample.

Many graphical and numerical techniques for representing and analyzing data are strictly valid only for measurements made on a ratio scale. It is important to remember this, because in digital computers all data are stored as binary numbers (bits), no matter what the scale of measurement.

One important characteristic of data in the earth sciences is that the place and time of measurement are important: data are spatially and temporarily distributed, and it rarely makes sense to ignore this fact. In other disciplines the place and time of measurement may not be important. For example, if we measure the density of pure water at a given temperature and pressure, we do not expect that it matters where the measurement is made or when. If a series of measurements give slightly different results, we assume that the differences are due to "random" errors of measurement. The best estimate of the "true" density may well be obtained by simply averaging all the measurements. But if we measure the density profile of water in a lake, the position of the samples in the profile matters and so does the position of the profile in the lake, (because we expect the density to vary with pressure, temperature, and salinity—which are all likely to vary systematically with depth and position in the lake), and the time of measurement is also important because we expect the density profiles in the lake to change from season to season, and perhaps to evolve slowly over longer time periods. So simply averaging all the measurements to obtain a representative density for the whole lake may not be a good idea.

1.3 Graphics and Descriptive Statistics

The word "statistics" has two quite different meanings. One is simply a number calculated from a sample of data. As such it can be considered simply as an attempt to describe some aspect of the data, and called a *descriptive statistic*. The other meaning is to describe the mathematical science of making inferences from numbers measured on samples (statistics in the first sense) about some larger "population" which we wish to study. This is using the word "statistics" to refer to the *science of statistics*. We give a very brief discussion of the science of statistics in the next chapter.

17.7	17.8	9.5	5.2	4.1	19.2	12.4	15.8
20.8	24.1	14.7	21.6	12.8	11.9	35.4	12.3
14.9	19.6	10.6	15.1	15.6	9.3	8.1	13.5
30.2	29.1	7.4	12.3	13.6	9.5	13.1	27.4
8.8	11.4	6.4	11.0	11.4	14.1	20.9	10.6
15.3	24.0	12.3	7.8	9.9	20.7	25.0	19.1
13.1	27.4	15.2	12.2	10.1	12.3	16.7	18.6
6.0	10.6	11.3	4.7	10.9	6.0	7.2	5.6
8.9	5.8	8.9	6.7	7.2	9.7	10.8	17.9
10.9	13.7	22.3	10.2	5.1	13.9	9.0	10.6
13.8	6.5	6.5	10.6	10.6	23.0	21.8	32.8
30.2	30.8	33.7	26.5	39.3	24.5	24.9	23.2
16.0	20.9	10.3	22.6	16.2	22.9	36.9	23.5
18.5	16.4	17.9	18.5	13.6	7.9	31.9	14.1
7.1	3.9	3.7	22.5	27.6	17.3		

Table 1.1   De Wijs data: Percent Zn in a sphalerite-quartz vein. The 118 assays at two meter intervals along the vein are ordered by row in this table.

Before considering the theoretical interpretation of statistics, we discuss the use of graphics and descriptive statistics to summarize data. This is actually a very large subject, because modern methods of remote sensing make it possible to collect very large numbers of numerical data, at almost every geophysical scale. Once scientists could afford to spend many hours pondering the meaning of small tables of numbers, but now it is imperative to have effective techniques for reducing the data to graphical or numerical summaries, before attempting interpretation. Computer techniques for "presentation graphics," "data visualization," and "geographic information systems" have been developed in recent years to meet this need. In what follows, we discuss only the most elementary methods, with some discussion of how they are implemented using MATLAB. All examples of MATLAB commands given in this book will be printed in typewriter font (see below). In this chapter, the MATLAB prompt will be shown as >>, though it may differ in the actual implementation available for your use. In later chapters, we omit the prompt.

First assume that the data consist of a series of measurements made, at equal intervals of time or space, of a single variable measured on the ratio or interval scale.

As an example, consider some classic data of De Wijs (1951): a series of 118 assays for Zn (zinc) made at two meter intervals along a single sphalerite quartz vein in the Pulacayo Mine in Chile (Table 1.1). From just looking at the data it is hard to see if there is any trend in zinc content along the vein. A better way to judge this is to plot zinc against distance along the vein. MATLAB provides several ways to do this. A file containing the data must first be prepared. In this case, the zinc values could be typed into an editor or word-processor, with a single number on each line, and the list saved as an ASCII file (see Appendix A). To save you the trouble, the diskette accompanying this book contains these data in the ASCII file `dewijs.dat`, together with all the other data and program files (M-files) discussed in this book. Suppose this diskette is in drive `B:`. After starting MATLAB, change the directory to this drive by typing

```
>> cd b:
```

and then load it into the program using the command:

```
>>  load dewijs.dat
```

The data are now in the computer memory, assigned to the vector `dewijs`. The data can be displayed using the MATLAB `plot` or `stem` commands. To plot the data using small circles for the data points (one of several options, see the MATLAB manual, or use `help plot`), enter

```
>> plot(dewijs, 'o')
```

To give the graph a title, and label the axes, for example, use the commands

```
>>  title('Zn assays in a Chilean vein');
>>  xlabel('Position (2m units)'), ylabel('Zn (%)')
```

If you wish to draw a line joining the points, you can do so in two different ways, either by holding the original graph on the screen and then plotting the lines, using

```
>>  hold on;
>>  plot(dewijs)
```

or by defining a vector (array of numbers) x with values going in increments of 2 from 0 to 234, and using this to plot both the points and the lines joining them with the commands

```
>>  x = [0:2:234]';
>>  plot(x, dewijs, 'o', x, dewijs)
```

Note that MATLAB vectors are always indexed from 1 to N, but the plotted x-scale may run between any limits: in this case we chose to begin at 0 meters, the spacing is 2 meters, and so the last sample is at $2(N - 1) = 234$ meters. Note also that to plot two sets of data against each other, they must be arranged in the same way. The zinc data was arranged as a single column of data (a *column vector*), so **x** must also be defined as a column vector. The command

```
>>  x = [0:2:234];
```

produces a single row of numbers, or *row vector*,

```
x = [0  2  4  6  8  10 ... 234]
```

but the prime (') converts this into a column vector.

The result (with a suitably adjusted `xlabel`) is shown in Figure 1.1.

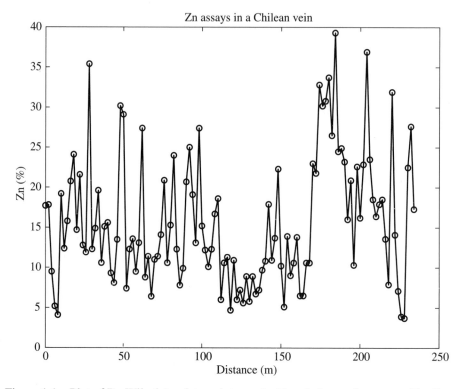

Figure 1.1   Plot of De Wijs data: data points marked by circles, and connected by lines

The Student Edition of MATLAB provides an alternative way of plotting data, which is useful for data measured at discrete intervals, if we have no reason to think that data at positions between the intervals either exist, or have intermediate values. This is the `stem` command

```
>> stem(dewijs)
```

The result is shown in Figure 1.2.

After examining these plots, we might conclude that there appears to be no systematic variation in Zn content along the vein. The variation appears to be erratic from one sample to the next (though we might note some tendency for values to be similar in adjacent samples). The next step might be to plot the data so that we can get a better impression of its *frequency distribution*, that is, its clustering about some central value (*central tendency*) and its spread about that value (*dispersion*).

To do this, we can use the MATLAB functions `hist` or `bar`. The terms *histogram* and *bar graph* are often used as though they were synonymous—but there is an important distinction between them. Both represent the frequency of the data in a series of classes (bins). For example, if we examine the De Wijs data we see that the frequency is 3 values in the range 0 – 4.9, 21 in the range 5.0 – 9.9, and so on. A bar graph represents the frequency

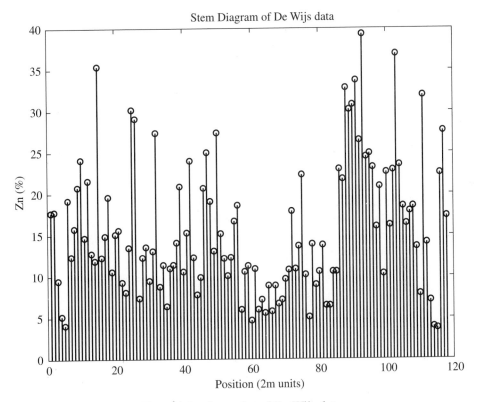

Figure 1.2   Stem plot of De Wijs data

by the height of a bar, whose midpoint is the midpoint of the interval (2.5, 7.5,... in our example). The width of the bar is not significant (and is usually the same for all intervals). In contrast, a true histogram represents frequency by the *area* of the bar. The width of the bar must be plotted to scale, and if the class interval changes, so do both the width and height of the bar.

Both `hist` and `bar` plot bar graphs rather than true histograms. The difference is not important so long as the class interval is constant. Using the command `hist(dewijs)` plots a bar graph of the De Wijs data using 10 classes whose midpoints are equally spaced between the highest and lowest values. Try this and examine the graph. You will see that some of the classes have a frequency less than 5. This is generally not advisable: too many classes for a small sample gives the bar graph a "ragged" appearance. A better graph can be obtained by specifying only 8 classes, using the command `hist(dewijs, 8)`. Better still, we may specify the exact class intervals to be used by first defining an array (vector) of class mid-points, for example

```
>> x = [2.5 7.5 12.5 17.5 22.5 27.5 32.5 37.5];
>> hist(dewijs,x)
```

If we want a plot that looks more like a conventional histogram, we can use the MATLAB function `stairs`. To do this we first use a variant of the `hist` function to obtain the vector of frequencies n, then use this vector to draw what MATLAB calls a "stairstep graph."

```
>> x = [2.5 7.5 12.5 17.5 22.5 27.5 32.5 37.5];
>> [n,x] = hist(dewijs,x);
>> x2 = [x-2.5 40];
>> n2 = [n 0];
>> stairs(x2,n2)
```

Note that we have to change from plotting midpoints (as in x) to plotting class limits (as in x2 – this involves subtracting a constant (2.5), and adding one more term to the vector). So we must also add a zero frequency to the end of the frequency vector (n2).

MATLAB Version 5 has an improved `hist`, but you may want to use `bar` to produce a version that allows easier control of the appearance, particularly if you want to print it in black-and-white. In Version 5 this can be done using

```
>> x = [2.5 7.5 12.5 17.5 22.5 27.5 32.5 37.5];
>> [n,x] = hist(dewijs,x);
>> bar(x,n,1,'w')
```

To understand the significance of the arguments in the bar function type `help bar`. The result is shown in Figure 1.3.

This series of MATLAB commands required to make a true histogram in Version 4 is sufficiently extensive that we may not want to have to type it in each time we want to plot a histogram in this way. So let us write a MATLAB function that we can save and use for many different data sets. Below, we simply list the function, well annotated with comments. The student should study the MATLAB manual to get a better understanding of this, our first, *user-defined function.*

```
function n = hist2(y, x)
% n = hist2(y,x)
% plots a histogram of the data in the vector y,
% using the vector of class limits given in x.
% The class intervals must all be equal, and the
% range of x should include all the data.  n is the
% vector of frequencies in each class.
% written by Gerry Middleton, September 1996.
ni = length(x) - 1;    % number of classes
dx = (x(2) - x(1))/2;  % find half the class interval
xmp = x + dx;          % determine class mid-points
xmp = xmp(1:ni);       % discard last mid-point
[n,xmp] = hist(y,xmp);
nn = [n 0];
stairs(x, nn)
```

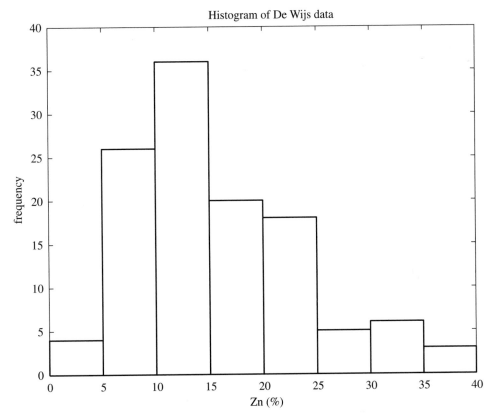

Figure 1.3   Histogram of the De Wijs data, using `bar`

Yet another way to represent data graphically is as a *cumulative curve* (Figure 1.4). This is a curve showing (on the ordinate) what frequency (or frequency percent) of the data is less than the value plotted on the abscissa. Cumulative curves are easily plotted in MATLAB using the `sort` and `plot` or `stairs` functions. First we sort the data in order of magnitude, then prepare a cumulative percent frequency scale ranging from 0 to 100%, and then plot:

```
>>  x = sort(dewijs);
>>  x = [0 x'];     % add zero and change to row vector
>>  y = 100*[0:118]/118;  % cumulative percent scale
>>  plot(x,y)
```

Cumulative curves are useful because no grouping of the data into arbitrary classes is required, as it was for plotting histograms. Also it is easy to determine *percentiles* from a cumulative curve, i.e., the percent of the sample that is smaller than some given value, or the value that corresponds to some percentile of interest (see Sec. 1.1).

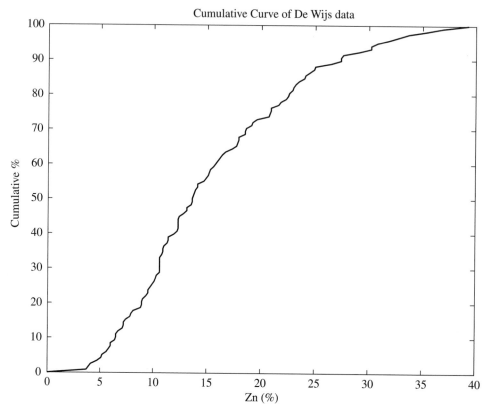

Figure 1.4   Cumulative Curve for De Wijs Data

Finally, we close this chapter by describing a few common descriptive statistics. The commonly used measures of *central tendency* are the *arithmetic mean*, the *mode* and the *median*. The mean of a sample of N values of x is given by the equation

$$\bar{x} = \sum_{i=1}^{N} x_i / N \tag{1.1}$$

In MATLAB it is easily calculated using the function mean. The mode is the class with the highest frequency. It can be determined approximately by inspecting the histogram. Some histograms show more than one maximum, i.e., the data are (or appear to be) *polymodal*. Spurious polymodality may arise from choosing too many classes to group a sample of limited size. The median is the 50 percentile, i.e., the value for which 50% of the sample is smaller (and 50% larger). It can be estimated graphically from the cumulative curve, or calculated from the sorted data. In MATLAB it can be obtained using the function median. The mean, mode and median are not necessarily the same value, unless the sample distribution is perfectly symmetrical. It is fairly clear that this is not the case for the De Wijs data. The center (12.5%) of the modal class (10–15%) is less than the median (13.65%) which in turn is less than the mean(15.61%). This is typical of data that is described as *skewed* to the smaller values, or having positive skewness.

The other commonly used statistics are those that describe the dispersion. These include the *range* of the data (the difference between the largest and smallest values), or various percentile ranges. For example, the *quartile range* is the difference between the 75 and 25 percentile. Any percentile may be estimated numerically using the function perc:

```
function xp = perc(x,p)
% xp = perc(x, p)
% return the pth percentile of a sample x
% written by Gerry Middleton, September 1996
n = length(x);
n1 = floor(n*p/100);   % index just below pth percentile
n2 = ceil(n*p/100);    % index just above pth percentile
x = sort(x);
xp = (x(n2) + x(n1))/2; % interpolated value
```

Better estimates of the dispersion are given by the calculated *variance* s^2 or *standard deviation* s, given by the equation

$$s^2 = \sum_{i=1}^{N}(x - \bar{x})^2/(N - 1) \tag{1.2}$$

These statistics are easily calculated using the MATLAB functions std or cov. For the De Wijs data $s = 8.01\%$. A crude estimate of the calculated value can be obtained from half the difference between the 84 and 16 percentile (8.08% for the De Wijs data). Further discussion of these important statistics is deferred to the next chapter.

1.4 Recommended Reading

In the references given in this book, the Library of Congress catalog number is given after the reference to a book.

Davis, John C., 1986, Statistics and Data Analysis in Geology. New York, John Wiley and Sons, second edition, 646 p. (QE48.8.D38 A well-written, comprehensive text, with many data sets and worked examples.)

Etter, D.M., 1993, Engineering Problem Solving with MATLAB. Englewood Cliffs, Prentice Hall, 434 p. (TA331.E88 Good supplement to the Tutorial in the Student Version.)

Griffiths, J.C., 1967, Scientific Method in the Analysis of Sediments. New York, McGraw-Hill, 508 p. (A pioneer work on statistics in the earth sciences: see Section 12.5 for a discussion of Stevens' scales of measurement.)

Stevens, S.S., 1946, On the theory of scales of measurement. Science, v.103, p.677–680 (The original definition of the four scales).

Chapter 2

Fundamentals of Statistics

2.1 Populations and Samples

When a scientist makes observations, for example, by measuring the thickness of beds in a stratigraphic section, it is generally *not* because he or she is exclusively interested in that particular set of data. Instead, the set of data is regarded as just one set that could be collected from a larger potential set of data sets, which is the real object of interest. In the example, that potential data set might be all sets of bed thickness that could (theoretically) be measured anywhere within a given stratigraphic unit. Statisticians call the object of interest the *population* and the set of data that have actually been measured the *sample*. Note that the population is a set of numbers, not a rock formation or some other natural object. The meaning of the numbers is defined largely by the operations used to measure them—so "bed thickness" is defined when the field geologist has a clear definition in his mind of how a "bed" is defined, and how to measure its thickness. Note also that the size of a population may be infinite—even if a stratigraphic unit has a finite areal extent and contains a finite number of beds, it is still possible (in theory) to measure their thickness at an infinite number of different sections.

It is further useful to distinguish between the *target* population, and the *sampled* population (Cochran et al., 1954; Krumbein and Graybill, 1965, p.147–169). Geologists may be interested in the whole stratigraphic unit, as it was originally deposited. But the unit can no longer be sampled because, since then, some (perhaps most) of it has been eroded away. Or the actual study may be limited to the existing outcrop of the unit because drill cores of the buried parts of the unit are not available. So the sampled population is necessarily quite different from the target population.

The science of statistics is about drawing inferences from samples about (sampled) populations. A scientist may make further inferences about the target population—but the reliability of these inferences cannot be quantified using the methods of the science of statistics. Numbers that describe a population are called *parameters* and are generally represented by Greek letters (e.g., the population mean and standard deviation are generally designated by the letters mu μ, and sigma σ). Numbers that are calculated from a sample are called *statistics*, and are generally designated by English italic letters (e.g., the sample mean and standard deviation are designated by "x bar" $\bar{x}$ and s). Note that a given population has only one value of a particular parameter, but a particular statistic calculated from different samples of the population has values that are generally different, both from each other, and from the parameter that the statistic is designed to estimate. Most of the science of statistics is concerned with how to draw reliable, quantifiable inferences from statistics about parameters. The subject is large, and only a sketch of some of the main concepts is given here.

2.2 Random sampling

Most people, even those who have taken a course on statistics, do not have a clear idea about what constitutes a random sample—yet the whole science of statistics is based on this concept, and most of the bad predictions made in the past (which gave rise to the phrase "lies, damned lies, and statistics") resulted from bad (i.e., non-random) sampling techniques. So it is very important to get clear what the word "random" means in this context.

A simple definition is that a random sample is one where every item in the population has an equal chance of being chosen. For example, if the population consists of just six items (the integers 1,2,... 6) then the technique of drawing a random sample must ensure that each item has an equal chance of being chosen. If we use a die to draw a sample, the die must be "true" (i.e., not weighted so that one number is more likely to appear than the others), and the method of casting the die must be fair. Where large sums of money are at stake, as they are in many public lotteries, choosing a number at random is not a trivial matter, and considerable ingenuity has been necessary to devise methods that are not only fair, but can be seen to be fair.

Suppose we want a sample of 10 numbers between 1 and 6. We can do this by casting a true die 10 times. We assume that a fair cast will ensure that, if a six was thrown at the last cast, it is no more or less likely to appear at the next one. In other words, the selection of every item in the sample is *independent* of every other selection. Notice that not many of the measurements made in the earth sciences can strictly qualify as independent observations. The concentration of Zn in specimens taken from a vein is likely to be similar to the concentration in specimens that are close by, and less similar to those that are far away. The discharge of a river measured in January 1996 is likely to be similar to the discharge in February of that year, and less similar to that in July.

Scientists may also have trouble defining what is meant by the phrase "equally likely selection." For example, in selecting sampling points on an outcrop consisting of six rock-types, should each rock type be sampled? Most geologists would agree that this might be a good procedure, but would not be a random sample of an outcrop, if for example, 95 percent of it was composed of one rock type. What if we want a random sample of the clasts from a gravel outcrop? At least three different sampling methods are possible: (i) select clasts so that each clast has an equal probability of being chosen (a very difficult task!); (ii) select clasts intersected by a line drawn on the surface of the outcrop—but these clasts must first have been intersected by the surface, and then by the line on the surface, so large clasts are more likely to be selected than small ones (in fact, the selection is weighted by the square of the clast size); or (iii) select clasts from points selected "at random" on the outcrop surface. Again the selection of large clasts is more probable—in this case, the selection is weighted by the cube of the clast size. If the clasts are collected to measure the size distribution, three different distributions would be obtained. If clasts were selected to measure the relative abundance of rock types, the abundances might also be different, if clast size was related to rock type (as it often is). In estimating rock type abundances, we are generally interested in volumetric abundances, so the third method would be the correct method to use.

Suppose we really wanted to collect a random sample of grains to measure their size distribution. Change the object to be sampled from an outcrop to a beach. Then we would have to devise a method that made it equally likely that each grain was chosen for measurement, and that each choice was independent of the others. This is obviously impracticable.

In practice, to estimate the size of grains on a beach, a sedimentologist collects a specimen of sand (each containing more than a million grains!) at systematically surveyed stations along and across the beach. The mean size of each specimen is determined by some mechanical process (such as sieving), and the mean size of the sample is determined by averaging the values determined at each of the stations.

This is a *systematic* sample, not a true random sample. The sampled population is really the set of all mean sizes that might be measured in this way, at all possible stations on the (surface of the) beach. Even using this definition, the sample can be considered random only at the local level, because at each station the choice of exactly where to collect the specimen is random, in the sense that the position is determined only to within a few meters. The local variation could be estimated by collecting pairs of specimens, chosen randomly at each station.

Left to themselves, earth scientists almost always choose systematic, rather than strictly random samples. We will see in later chapters why this generally makes good sense. Most data in the earth sciences are distributed in space or time. Data are collected at regular intervals because that makes it easier to evaluate spatial or temporal trends. If we can use the data to show that such trends do not exist, then the regularity of the sampling does not matter: nature's randomness compensates for the scientist's regularity, and the sample may be treated as "essentially random."

We must, nevertheless, be on guard against errors in interpretation. For example, if a sequence of temperature measurements are collected at a weather station every day at midday, averaging them over a year provides an estimate of the mean midday temperature during that year, but not an estimate of the average yearly temperature (because generally the temperature during the day is higher than at night). This is an example not only of a different result produced by systematic and random sampling, but also of the importance of distinguishing between the target population and the sampled population.

MATLAB can help us generate some random numbers, if we need them. In the old days, statisticians who needed random numbers would turn to a table of random numbers in a text (this table was, in its turn, generated by a computer). MATLAB has two functions (`rand` and `randn`) that generate numbers that are "very nearly random" (*pseudo-random* in the cautious jargon of modern computer science). The first (rand) consists of numbers between 0 and 1, drawn from a *uniformly distributed population*; that is a population with an equal frequency of values in any two intervals of equal size, e.g., numbers in the range 0.1 – 0.1999 have the same frequency as numbers in the range 0.2 – 0.2999. The second (randn) consists of numbers drawn at random from a Normal population with mean 0 and variance 1 (see Section 2.5).

Suppose we want to draw three samples at random from the 118 samples in the De Wijs data. Each sample in the array is identified by its row number, so the problem is to draw three integers at random in the range 1 to 118. We could do this using MATLAB as follows:

```
n = ceil(rand(3,1)*118)
```

rand(3,1) produces a column of 3 random numbers between 0 and 1, multiplying this by 118 converts them into three real numbers between 0 and 118, and the function ceil rounds them upwards to integers between 1 and 118.

Note that it is quite possible that two of the integers might be the same. This is "sampling with replacement." If we do not want this, but rather "sampling without replacement," two procedures are open to us: (i) we could simply reject the sample and try again, or (ii) we could use MATLAB's randperm. Use of randperm(118) generates a random permutation of the integers between 1 and 118: we can then use the first three of these to identify three random samples of the De Wijs data.

2.3 Probability

In the past, the term *probability* has been defined in several ways. Today, it is generally defined axiomatically, using concepts from set and measure theory. Though we will not attempt to give a full presentation of this definition, we give an outline and illustrate how the abstract concepts used apply to the example of casting dice.

First define a "sample space", consisting of the set of all possible outcomes of the operation being considered. For a single die this is the set (1,2,3,4,5,6): i.e., in our ideal world, the result of casting a die must be one of these numbers—we do not consider the possibility of the die landing on its edge, or of losing it.

Next define an "event" as any subset of the sample space. For example, a possible event for throwing a die once is one of the numbers 1,2,...,6. A "random event" is one where all the events are equally likely.

Finally we define the probability P of a random event, as being $1/N$, where N is the number of events in the sample space. For the throw of a single die: there are 6 equally likely "elementary events" in the sample space, so the probability of obtaining any one of them is 1/6. Two events are said to be mutually exclusive if they cannot both occur, and an event A consisting of at least one of two such events A_1 and A_2 is said to be their "union", written $A = A_1 \cup A_2$. So in casting a die, getting either a 1 or a 2, is an event consisting of the union of two mutually exclusive events 1 and 2. The probability of this event is defined to be the sum of their separate probabilities (i.e., $(1/6) + (1/6) = 1/3$). The only event that is certain, is the union of all the elementary events, for example, getting one of the six numbers in the sample space for a die, and it has a probability of one. Impossible events are those that are not in the sample space (e.g., 7) and they have a probability zero.

It turns out that all of probability theory can be deduced from three axioms:

1. The probability P of an event A, must lie between 0 and 1, or

$$0 \leq P(A) \leq 1. \tag{2.1}$$

2. For the set of all elementary events P is one.

3. The probability of the union of two independent events is the sum of their probabilities, or

$$P(A_1 \cup A_2) = P(A_1) + P(A_2) \tag{2.2}$$

These axioms were first formulated rigorously by the Russian mathematician, A.N. Kolmogoroff, one of the great mathematicians of the 20th century, who made many other contributions to pure and applied mathematics, including a theory of turbulence at large scales, and a theory for size distributions of sand. He has given a very readable account of the axiomatic theory of probability (Kolmogoroff, 1963). Alternatively, for a modern introductory text that gives a good discussion of probability as well as statistics see Berry and Lindgren (1996).

Of course, not all problems in probability can be immediately solved using the axioms, but all the fundamental properties of probability can be deduced from them. For example, suppose there are two dice, X and Y, thrown simultaneously. The total set of elementary events would be the set of ordered pairs (1,1), (1,2), (1,3), (1,4,), (1,5), (1,6), (2,1) and so on (the numbers are the value of dice X and Y, respectively). There are 36 such events, so the probability of any one is 1/36. A total score of 3, however, could be obtained in two ways, (1,2) and (2,1), so its probability is the sum of the two probabilities of the elementary events, or 1/18. What if we throw the two dice one after the other? If the first one (X) is 1, what is the probability of a 2 on the throw of the second one Y? If the answer is 1/6, then we say that the two events are independent. It then follows that the probability of a combination of two independent events is equal to the product of their individual probabilities, which is 1/36—not coincidentally, this is the same as when we threw the two dice together.

Probability theory gets more interesting, and even controversial, when it deals with events that are *not* independent. We then define the *conditional probability* $P(A|B)$ of event A, given that B has already occurred, as

$$P(A|B) = \frac{P(AB)}{P(B)} \tag{2.3}$$

where $P(AB)$ means the probability that both A and B occur. Conditional probabilities can be calculated from *Bayes' Theorem*, which states that

$$P(B|A) = \frac{P(A|B)P(B)}{P(A)} \tag{2.4}$$

Bayes Theorem may be proved from the axioms and the definition of conditional probability. Controversy arises when information obtained by an investigator (sometimes just from general experience) is used to estimate B, before a sampling experiment is carried out. This introduces a "subjective" element to the application of statistical methods: leading to an approach often called *Bayesian statistics* to distinguish it from classical statistics. For an elementary treatment of Bayesian statistics see Berry (1996).

The following example illustrates a straightforward application of Bayes' theorem. Floods are often caused by spring melting of snow. They are more probable when there has been a heavy accumulation of snow in the winter. In a given drainage basin suppose that a long series of observations has established the probability of a flood $P(F)$ to be 0.3, and of a heavy snow accumulation $P(S)$ to be 0.15. It is further known that the probability of a flood, following a heavy snow accumulation $P(F|S)$ is 0.6. What is the probability that any given flood was preceded by a heavy snow accumulation? Bayes' theorem gives the answer:

$$P(S|F) = P(F|S)P(S)/P(F) = (0.6 \times 0.15)/0.3 = 0.3$$

What this tells us, is that although heavy snows generally produce floods, most floods are not produced by heavy snows in this (hypothetical) basin.

Statistical methods are often applied by scientists in order to make decisions about accepting or rejecting hypothesis more "objective." Applying statistical tests, based on probability theory, may indeed quantify concepts such as "the best estimate of the mean" (see the discussion later in this chapter). But it is well to remember that the probabilities of real events are never known with the certainty assumed of mathematical events. The probability of obtaining a 6 by rolling a real die is not necessarily 1/6, because a real die does not behave exactly in the way the mathematical model assumes it does. Supposing we threw a real die 100 times, and obtained a six 50 times. Would we still believe that the die was "true," or would we change our estimate of the probability of getting a six on the next throw, based on the hypothesis that the die was loaded? In the real world, full objectivity remains an elusive goal.

A common concept of probability, now generally thought to be inadequate for a complete theory, is that it is the relative frequency of occurrence of the event "in the long run." If we throw a true die 1000 times (and believe it or not, there was once a time when people tried to establish an experimental basis for probability theory by doing just that), then we expect that about 1/6th of the events will be ones, and so on. The larger the number of throws, the closer we expect the ratio will approach 1/6. This concept of probability is often used in explaining the results of polls: for example, it is said that the results of a poll indicate that "on the average, 19 out of 20 such polls would show that between 50 and 60 percent of respondents will choose brand X over brand Y."

2.4 Bias, Consistency, and Efficiency

A more exact definition of the value that a statistic will have "in the long run" is given by the concept of statistical *expectation*. We first define a *random variable* as a variable x, which in different sampling experiments, assumes different values x_i, each of which is a random event (as defined in the previous section). Note that some data, such as the numbers obtained from a die, are inherently *discrete*, that is they can take only integral values. Other data, such as measurements of length, are *continuous*, because they can take decimal values. Random variables can be either discrete or continuous, but somewhat different mathematical techniques apply to each type.

The *expected value* or *expectation* $E(x)$ of x is defined to be

- for a discrete variable,

$$E(x) = \sum_i p_i x_i \qquad (2.5)$$

  where p_i is the probability of the ith value x_i of x.

- for a continuous variable, where the probability in some infinitesimal interval dx is expressed as a *probability density function* $\phi(x)$ (see the next section for an example)

$$E(x) = \int x\phi(x)\, dx \qquad (2.6)$$

The expected value of throwing a die is therefore the sum

$$E = 1/6 + 2/6 + 3/6 + 4/6 + 5/6 + 6/6 = 21/6 = 3.5$$

This is the value we expect to get if we compute the mean value of throwing a die, averaged over a large number of throws. Note that it is not an integer, because averages of integers are generally not integers.

If the expected value of a statistic is equal to the value of a population parameter, then we say that the statistic is an *unbiased estimator* of that parameter. For example, if $E(\bar{x}) = \mu$ (the expectation of the sample mean is equal to the population mean), then the sample mean is an unbiased estimator of the population mean. Lack of bias is obviously a very desirable property when we use statistics not just to describe a sample, but to estimate a population parameter, but it is not the only desirable property.

Another property of statistics is *consistency*. Suppose the parameter is α, estimated by a statistic a, based on a sample of size N (i.e., computed from N independent random single sample-events). Then if $a \to \alpha$ as $N \to \infty$ (i.e., if a approaches α ever more closely as the sample size increases) the estimator is said to be consistent. The difference between bias and consistency is that bias deals with averages of statistics calculated from a large number of (finite sized) samples, while consistency deals with the effect on a statistic of increasing sample size.

An important example is the following: if the population is Normally distributed (see below) then it can be shown that the statistic

$$s^2 = \sum (x_i - \bar{x})^2 / N$$

is consistent, but biased for small samples. This is why we use the definition of variance given in the previous chapter, where the denominator is $N - 1$, not N.

Finally, a statistic might be both unbiased and consistent, yet fail to be *efficient*. An example (published in verse, as a parody of Longfellow's Hiawatha!) was given by the great statistician M.G. Kendall (1959): Hiawatha shoots arrows at a target, and never hits it—but he argues that his arrows miss the target on every side, so that if you computed an average position, it would be at the bullseye. His shooting therefore is unbiased and consistent. Other archers, however, manage to hit the target, though their arrows cluster on a small area of the target, just away from the bullseye. Whose shooting is better?

A statistic a is said to be an efficient estimator of α if the variance of a is smaller than that of any other estimator, for a given sample size. The ability of an archer to cluster his arrows on a small part of the target is an example of efficiency. If the cluster coincides with the bullseye, then the shooting is also unbiased, otherwise it is biased.

The statistical terms unbiased and efficient correspond closely to two terms used in analytical sciences: *accuracy* and *precision*. An analytical method is accurate if the average of several replicate analyses lies close to the "true" value (as determined, for example, for a well known standard such as the granite G1 or basalt W1 standards set up by Fairbairn, 1951). It is precise if the standard deviation of several replicates is small. Both are desirable properties, but precision is probably more important than accuracy, because if the bias can be determined (by analyzing a standard) then the analyses can be corrected accordingly. But if the precision is low, the only safeguard is to make many replicate analyses, and average them.

Modal analysis by point counting is a technique that estimates the volumetric proportion of a mineral in a rock by counting the number of times grains of the mineral underlie the intersections of a regular overlain grid on a thin section or polished surface. It can also be used to estimate the areal proportion of features (such as lakes) on maps. It can readily be applied by computers to the analysis of digitized images, since all that is required is to count the number of pixels having a given color or shade. The technique is now well established in the earth sciences, but it was originally viewed with suspicion until a careful theoretical and experimental analysis by Chayes (1956) proved that it was both accurate and precise, and gave better results than other ways of estimating mineral compositions. It also provides an example of the good use of a systematic sampling technique, rather than one which is totally random.

2.5 The Normal Distribution

A common assumption made by statisticians, is that a population has a *Normal* or *Gaussian* distribution. Most statistical tests (such as the t-test and F-test described later in this book) are based on this assumption. In this section, we consider what assuming a Normal distribution means, why such an assumption is necessary, and why we need not worry too much about whether or not it is exactly valid.

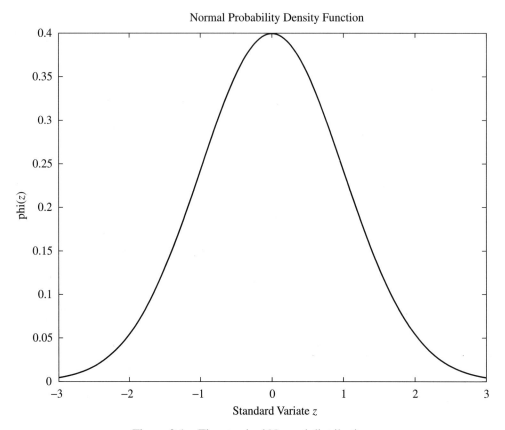

Figure 2.1   The standard Normal distribution

The Normal distribution is an example of a *probability density function*. It is a bell-shaped curve (Figure 2.1) of a function $\phi(z)$

$$\phi(z) = \frac{1}{\sqrt{2\pi}} e^{(-z^2/2)} \tag{2.7}$$

where z is the "standard variate"

$$z = \frac{x - \mu}{\sigma}$$

x is a random variable, and μ and σ are the population mean and standard deviation.

The curve is called a probability density function because (i) z (and x) is a continuous random variable, (ii) the total area under the curve is one, (iii) the area to the left of any ordinate z is the probability of obtaining a value less than or equal to z by taking a single random sample of the population. The curve is symmetrical, so the mean is equal to the mode and also to the median, and has the value $z = 0$ (or $x = \mu$). The population variance

is defined as the expected value of the mean square deviation (z^2 in this case). Using
Equation 2.6, it can be shown that, for standard variates, it has the value $\sigma_z^2 = 1$ (or for x,
$\sigma_x^2 = \sigma^2$). The inflection points on either side of the mean lie at a distance of one standard
deviation from the mean. The corresponding ordinates ($z = -1$, $z = +1$) enclose an area
of 68%, and so these values correspond to the 16 and 84 percentiles of the distribution. The
values of $\phi(z)$ (the ordinates of the Normal probability density function) or the cumulative
probabilities (i.e., the areas under the curve left of any ordinate z), given by

$$p(z) = \int_{-\infty}^{z} \phi(z)\, dz \tag{2.8}$$

are given in any book of mathematical or statistical tables.

The integral shown above cannot be solved analytically, but it may be approximated by
numerical techniques. MATLAB's function `erf` calculates the *error function* often used in
applied mathematics: It is defined by the equation

$$\mathrm{erf}(z) = \frac{2}{\sqrt{\pi}} \int_{0}^{z} e^{-z^2}\, dz \tag{2.9}$$

Thus $\mathrm{erf}(z)$ is not quite the same thing as the cumulative probability distribution $p(z)$: it
gives twice the integral of the Normal distribution from 0 to $z/\sqrt{2}$, rather than the integral
from $-\infty$ to z. (Note the scaling of the variable, which is not explained in the MATLAB
handbook.) For example $\mathrm{erf}(0) = 0$, $\mathrm{erf}(1/\sqrt{2}) = 0.64$. So the following function
computes the cumulative probability:

```
function p = cump(z)
% p = cump(z)
% computes the cumulative probability for a normal
% distribution with standardized variate z
p = 0.5 + 0.5*erf(z/sqrt(2))
```

MATLAB also provides a useful function `erfinv` that returns the inverse of the error
function. This allows us to compute the ordinate z that corresponds to a probability p, if
we scale the result as shown in the following function:

```
function z = cumpinv(p)
% z = cumpinv(p)
% computes the standard ordinate z, of a Normal distribution
% corresponding to a cumulative probability p
z = erfinv(2*p - 1)*sqrt(2)
```

To plot the normal probability function for a set of z-values (e.g., `z = [-3:0.01:3]`),
we can use the following function (this function was used to produce Figure 2.1):

```
function dp = normplot(z)
% dp = normplot(z)
% computes the probability density function of a Normal
% distribution at the standardized variates in the
% vector z, and plots it
% written by Gerry Middleton, September 1996
n = length(z);        % no of variates
r = max(z) - min(z);  % range of variates
pc = 0.5 + 0.5*erf(z/sqrt(2));  % cumulative probabilities
pc2 = pc(2:n);
dp = (pc2 - pc(1:n-1))*n/r; % probabilities for z increments
zmid = z(1:n-1) + r/(2*(n-1)); % midpoints of z increments
plot(zmid, dp)
```

Note that in plotting a probability density function, we have to scale the ordinate so that the total area under the curve is equal to the total probability, i.e., one. To do this numerically, we divide the total range into a large number of class intervals (e.g., z = [-3:0.06:3]), find the cumulative probability at each class boundary, calculate the probability of each class as the difference between cumulative probabilities at the lower and upper class boundaries, and plot this probability at the midpoint of the class.

The importance of the Normal distribution in science is a consequence of three facts, one observational and the other two theoretical;

1. Much "natural variation" is observed to be approximately Normally distributed. This applies particularly to errors of measurement or chemical analysis. Thus the Normal distribution serves as a useful "model" of the (unknown) true distribution.

2. The Normal distribution is simpler mathematically than many alternative models. For example, a particular Normal distribution is completely defined by only two parameters (the mean and standard deviation). Rival distributions (e.g., the "exponential" distribution championed by Barndorff-Nielsen, see Barndorff-Nielsen and Christensen, 1988) generally require more parameters, and the theory for their application has not yet been developed fully.

3. A part of statistical theory (generally described rather imprecisely as the "central limit theorem") indicates that averages of samples taken from a (non-Normal) population tend to follow a Normal distribution much more closely than the original population (Figure 2.2).

The central limit theorem provides a basis for a theoretical model of errors of measurement. Suppose a single measurement can be considered to be the sum of the "true value" x_t, and a large number of small, independent errors $\epsilon_1, \epsilon_2, \ldots$.

$$x = x_t + \epsilon_1 + \epsilon_2 + \ldots + \epsilon_n. \tag{2.10}$$

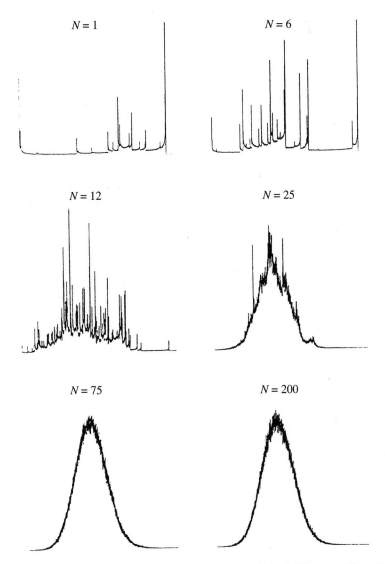

Figure 2.2 Computational demonstration of the Central Limit Theorem (from Eubank and Farmer, 1990). The frequency distribution of the original data is shown as $N = 1$: it is highly polymodal. The other curves show the distribution of averages of different sample size N. Even for these data, a close approximation to the Normal distribution is shown for sample sizes larger than 50.

Then the central limit theorem tells us that if n is large, x will be approximately Normally distributed. This result was derived by the mathematician C.F. Gauss, which is why the Normal distribution is also called the Gaussian distribution.

The *Lognormal distribution* can be considered to be a variant of the Normal distribution. In this case it is not the values x themselves that are Normally distributed, but their logarithms. If a histogram of observed x_i is examined, it can be seen that it is not symmetrical, but skewed towards the smaller values. Such histograms are typical of much observational data, so it is often worth trying a logarithmic transformation to see if this produces a more symmetrical distribution. In some areas of research, study of natural grain size distributions, such a transformation has become part of the standard grain-size scale, called by sedimentologists, the *Phi scale*:

$$\phi = -\log_2 d$$

where d is the grain size, measured in millimeters. The negative sign is introduced so that the smaller sizes become positive.

The central limit theorem can be used to provide an explanation for the Lognormal distribution as well as the Normal distribution. We simply suppose that the small errors (or natural sources of variation, giving rise to the distribution) are proportional to the magnitude of the true value, that is they are small multiples (proportions) of that value, not additions to the value:

$$x = x_t \times \epsilon_1 \times \epsilon_2 \times \ldots \times \epsilon_n. \tag{2.11}$$

If we take logarithms of this equation, it is reduced to the same form as Equation 2.10, so we expect that $\log x$ will be Normally distributed. Why should "errors" be proportional to the magnitude of the value being measured? We can see why by considering size: it is obviously much more difficult to measure a meter-wide boulder to within one tenth of a millimeter, than it is to measure a millimeter-wide sand grain to the same accuracy. The same "proportionality effect" also applies to many sources of natural variation—large flood currents or waves, capable of moving boulders, show a much larger variation in absolute magnitude, than the lesser currents or waves needed to move sand. In other words, variation should be assessed relative to the magnitude of what is varying rather than in absolute terms. This is why it is often expressed as the *Coefficient of Variation* $s/\bar{x}$, rather than as the standard deviation.

2.6 Are Data Normally Distributed?

Often we have collected a sample of data, and wish to know if the sample might have come from a Normal population. Here we consider two graphical approaches to this problem, and in the next section we consider a more sophisticated numerical approach.

The first approach is to compute the best fit Normal distribution, and compare it with a histogram of the data. As we have seen, the Normal distribution has only two parameters, the mean and standard deviation, which are best estimated using the sample mean, and sample standard deviation. Once we have these two estimates we can draw a Normal distribution, but instead of drawing the standardized version (z versus $\phi(z)$), we draw a version where the non-standardized variable x is plotted, and the area under the curve is equal to the total frequency times the histogram class interval. The details of how this is done can be seen by

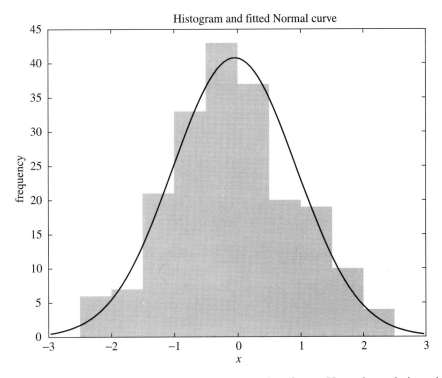

Figure 2.3   A histogram of 200 values drawn at random from a Normal population with mean 0 and standard deviation 1, together with the fitted Normal curve.

examining the MATLAB script normfit: an example of the result, where the data consists of 200 values from a Normal distribution produced using the MATLAB function randn is shown in Figure 2.3.

```
function [f, chi2] = normfit(y, x, c)
% [f, chi2] = normfit(y,x,c)
% plots a histogram of the data in the vector y,
% using the vector of class limits given in x.
% The class intervals must all be equal, and the
% range of x should include most of the data.  n is the
% vector of frequencies in each class.  Also plots
% the fitted normal distribution, and calculates ChiSquared
% If c = 1, then the histogram will be color-filled
% written by Gerry Middleton, September 1996.
ni = length(x) - 1;    % number of classes
N = length(y);         % number of data
dx = (x(2) - x(1))/2;  % find half the class interval
xmp = x + dx;          % determine class mid-points
```

```
xmp = xmp(1:ni);          % discard last mid-point
[f,xmp] = hist(y,xmp);
nn = [f 0];
if c == 1                 % for filled histogram
    cc = [f;f];           % ff and xx give coordinates of polygon
    cc = reshape(cc,1,2*ni);
    cc = [0 cc 0];
    xx = [x;x];
    xx = reshape(xx,1,(2*ni + 2));
    fill(xx,cc,'g');      % change 'g' to another color if preferred
    line(xx,cc);
    hold on;
else                      % for unfilled histogram
    stairs(x, nn);
    hold on;
    plot([x(1) x(1)], [0 nn(1)]);
end
hold on;
m = mean(y);              % mean value
s = std(y);               % standard deviation
p = 0.5 + 0.5*erf((x-m)/(sqrt(2)*s));  % cum norm prob of class limits
pt = p(ni+1) - p(1);      % total probability in classes
p2 = p(2:ni+1);
pc = p2 - p(1:ni);        % prob of each class
fc = pc*N*pt;             % theoretical frequency in each class
chi2 = sum((f - fc).^2./fc); % calculate Chi Square
dx2 = (max(x) - min(x))/100; % now plot Normal curve
x2 = [min(x):dx2:max(x)];
x2m = x2(1:100) + dx2/2; % 100 midpoints
p3 = 0.5 + 0.5*erf((x2-m)/(sqrt(2)*s));
p3 = p3(2:101) - p3(1:100);
fc2 = p3*pt*N*100/ni;
plot(x2m,fc2,'r','LineWidth',1);
title('Histogram and fitted Normal curve');
xlabel('x'), ylabel('frequency');
hold off
```

The first part of this function is very similar to hist2. The second part is similar to normplot but uses the calculated mean and standard deviation of the data, and the sample size, to calculate the frequencies in each class for plotting instead of the probabilities. The scaling is such that the total area under the curve is the same as the total area of the plotted histogram. The function also calculates the *Chi Square* statistic for the fit: the use of this statistic is explained in Section 2.7. In Chapter 3, we will present a revised version of this function.

The second approach is to plot the cumulative curve on a special type of graph paper, called *probability paper* which has an ordinate scale that converts a Normal cumulative curve from a sigmoid curve to a straight line. We can then estimate by eye whether or not the data show a close approximation to a Normal distribution. We expect that the central part of the distribution should show a better approximation to a straight line than the "tails" (lowest and highest values), because of the low frequencies represented by the tails.

Probability paper can be purchased from any university bookstore, but it is easier to use the MATLAB function cumprob:

```
function [m, s] = cumprob(x)
% [m, s] = cumprob(x)
% plots the data in the vector x as a cumulative
% curve on probability paper: also plots the best fit
% Normal, calculated from the mean m and standard deviation s
% The circle indicates the mean (with a two sd error bar)
% and the line extends from -2*s to + 2*s.
% written by Gerry Middleton, September 1996
m = mean(x);
s = std(x);
xf = [m-2*s, m+2*s];    % calculate the values 2 std from the mean
yf = [-2 2];
n = length(x);
e = 2*s/sqrt(n);    % estimated error of mean
x2 = sort(x);    % sort data
y = cumsum(x2);    % calculate cumulative sum
x2 = x2(1:n-1);    % remove largest value
y = y(1:n-1)/y(n);    % standardize cumulative sum
y2 = sqrt(2)*erfinv(2*y - 1); % convert to Normal probabilities
% Set up cumulative Normal probabilities y3 corresponding to
% chosen percentiles y4, for ordinate of graph
y3 = [-2.3263 -1.2816 -0.8416 -0.5244 -0.2533 0.0 ...
            0.2533 0.5244 0.8416 1.2816 2.3363];
y4 = [' 1';'10';'20';'30';'40';'50';'60';'70';'80';'90';'99'];
plot(x2,y2, xf,yf, m, 0, 'o');
% change default ordinate ticks and labels to required percentiles
set(gca, 'YTick', y3, 'YTickLabels', y4)
title('Cumulative Curve, on Probability Paper');
xlabel('x'), ylabel('cum. freq. (%)')
hold on;
% add the mean, with error bar, and best fit Normal line
plot([m-e,m+e],[0 0]), grid on
```

This script is a little more complicated than others that we have listed so far, but much of it is concerned with graphics details. The crucial computation is performed by the MATLAB function erfinv, which we use to convert the standardized cumulative frequency scale to

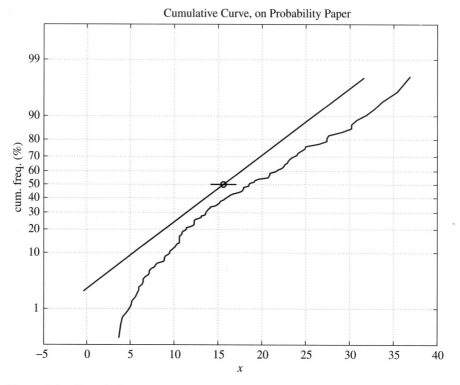

Figure 2.4   Cumulative curve of the De Wijs data, plotted on probability paper. The mean is shown by the circle, and the straight line is a plot of the best fit Normal, over a range of $4s$.

a scale of deviations from the mean, measured in z units. y3 is a vector containing the z values that correspond for a Normal distribution to the vector of cumulative percentiles listed in y4.

Figure 2.4 is a plot of the De Wijs data using cumprob. It is clear that these data show substantial departures from a Normal distribution. Note too that the Normal "fit" obtained from calculating the mean and standard deviation is quite different from the line that would be obtained by fitting a straight line to the central region of the cumulative curve.

2.7   Sampling Distributions

2.7.1   Comparing Observed and Predicted Frequencies

Suppose we have a theoretical model of a population that we can use to predict the frequencies by class in a random sample. For example, if we assume a true die, we predict that in a random sample of size N, each number will have a frequency of $N/6$. Is there a theory that tells us not only the predicted value, but also the expected magnitude of the observed deviations from the theory? Such a theory was worked out by the pioneer mathematical statistician Karl Pearson. He found that, for samples drawn from many types of population, and classified into n classes, the statistic

$$\chi^2 = \sum_i^n \frac{(f_i - \hat{f}_i)^2}{\hat{f}_i}$$

was distributed as a distribution now known as the *Chi Square* distribution. The f_i are the observed frequencies, and the $\hat{f}_i$ are the predicted frequencies. The Chi Square distribution has one parameter, called the "degrees of freedom." This is the number of independent observations needed to calculate the statistic. In this case the observations are the six frequencies, but they are not all independent because there is one "constraint:" they must add up to the total sample size N. No other theoretical constraints are necessary, so the total degrees of freedom (d.f.) are 5.

The Chi Square distribution predicts the probability of obtaining a value of χ^2 of any given magnitude, for a given d.f. Its value is tabulated in books of mathematical or statistical tables. For example, for 5 d.f., the probability of getting $\chi^2 > 15$ is about 0.01. This is a low probability, so if we observed such a high value on an experimental test with a die, we might well suspect that the die is not true, or was not being fairly thrown.

Another example of the use of Chi Square is to test the "goodness-of-fit" of an observed distribution to a Normal distribution. Suppose we have N observations, classified into n classes. We have seen that we can use the Normal cumulative distribution to calculate a theoretical frequency for each class. So we can calculate χ^2. But in this example there are three constraints: not only must the frequencies add up to N, but we can only calculate their theoretical values after we have first calculated the sample mean and standard deviation (the two statistics used to estimate the best-fit Normal distribution). So the degrees of freedom are $(n - 3)$. Figure 2.3 showed an example of a calculated fit to random Normal data: there are 12 classes, so the degrees of freedom are 9. The calculated Chi Square was about 8, and tabulated values show this is well below the value for a probability of 0.01 (21.7): so we accept the hypothesis that these data were drawn at random from a Normal distribution.

As a second application of Chi Square to test goodness of fit, consider Figure 2.5. This shows the result of applying the function `normfit` to the De Wijs data. The fit is clearly not good (as we also saw in the plot on probability paper, Figure 2.4): but might this be due to random sampling? The number of classes is 8, so the degrees of freedom are 5, and the calculated Chi Square is 23.4. This is substantially larger than the tabulated value for a probability of 0.01 (15.1): so we conclude that the original population was not Normal.

These applications of Chi Square are our first examples of *statistical testing*. The steps used in applying a statistical test are as follows:

- Make a *null hypothesis*. In our two examples, the null hypotheses were (i) the die was a true die, and (ii) the population was a Normal distribution. We must make a null hypothesis, because without it we have no way of predicting the probability distribution of the calculated statistic. In our examples, without a null hypothesis we

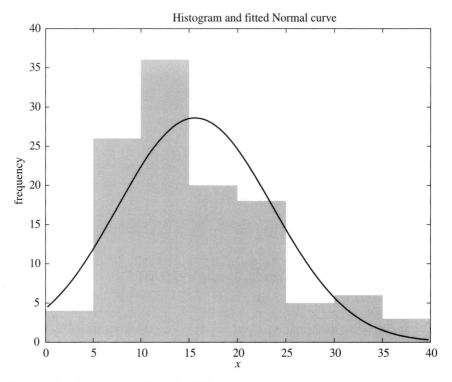

Figure 2.5   A histogram of the De Wijs data, together with the fitted Normal curve.

could not predict the distribution would be Chi Squared, or know the correct number of degrees of freedom.

- Set a *significance level*. This is the (low) probability α at which we will reject the null hypothesis. The conventional choices are 0.05 or 0.01 (5% or 1%).

- Calculate the statistic and its degrees of freedom.

- Look up its predicted value in mathematical tables for the appropriate degrees of freedom, and chosen significance level.

- If the observed statistic is larger than the tabulated value we reject the null hypothesis.

In the Chi Square examples presented above, we should note some reservations: (i) The smallest theoretical frequency for any class should not be less than 5. This was not the case in the two examples presented, so we should really recalculate Chi Square with fewer classes or a smaller range, for example, using x = [0:5:35] for the De Wijs data: the result is that the two largest classes are grouped together. (try this—you will find it does not seriously affect the result in these two cases). (ii) We noted in Chapter 1 that the De Wijs data does not constitute a true random sample, but rather a systematic sample of the vein. Perhaps a true random sample would not show the same deviations from a Normal distribution. Plotting the data, however, seemed to show that there was no clear trend in zinc values along the vein: in this case, we may argue that Nature's randomness converts a systematic sample into one that is "almost random."

2.7.2 Statistics of Samples from a Normal Population

In this section we consider tests that apply strictly only if we are prepared to assume that the population has a Normal distribution. In many cases, however, these tests are applied even though it is not really known whether this assumption is valid. Many studies have shown that small departures from exact population Normality do not significantly affect the results of these tests, and the tests retain some validity even if there are very large deviations from Normality, such as a strongly bimodal population. In such cases, the theoretical significance level may be in error by a factor of 2 or more, but the results still have some qualitative significance.

If the population is Normal, then the purpose of sampling is generally to estimate the population mean or variance. Hypotheses about the mean may be tested using the Normal distribution (for large samples) or the Student's t-distribution for small samples. Hypotheses about the variance may be tested using the Chi Square or F-distribution.

For large samples from a Normal population, we may be willing to accept that the sample variance is a sufficiently accurate estimate of the population variance (or we may know the population variance from some other studies, for example, the error variance of an analytical technique may be known accurately, because of a large number of analyses of standard specimens). In this case, it is known that the means of samples of size N will themselves be Normally distributed, with variance σ^2/N. We can therefore predict that 95% of all the sample means will lie within a distance of two (more precisely, 1.96) standard deviations of the population mean. So if we make a null hypothesis, that the population mean has the value μ_0, we can predict that 95% of all samples of size N should lie in the range $\mu_0 \pm 1.96\sigma/\sqrt{N}$. If the observed value lies outside this range, then we reject the null hypothesis at the $\alpha = 0.05$ level of significance.

A common practice is to use the information about the variance of the mean, not to test a hypothesis about the population mean, but to set up *confidence limits* for the observed sample mean. If a null hypothesis can be used to predict the range of variation expected in sample means, then the same range about the observed mean shows which null hypotheses about the mean can be accepted: we accept any null hypothesis about the population mean that lies in the range $\bar{x} \pm 1.96\sigma/\sqrt{N}$. In words, this is expressed by saying that these are the 95% confidence limits for the observed mean, or that we are 95% confident that the population mean lies somewhere in this range. Testing null hypotheses and setting confidence limits are really just two possible interpretations, based on the same data and assumptions about the population it came from.

The *Student's t-distribution* can be used to test hypotheses about population means, or set confidence limits for sample means, in cases where the variance must be estimated from the sample itself. In practice, the difference between the results of a t-test and a test based on the Normal distribution is only important for small samples ($N < 25$). For small samples, the 95% confidence limits are given by $\bar{x} \pm t_{0.05}s_{\bar{x}}$, where the subscript 0.05 indicates the α level, and the $s_{\bar{x}}$ is the estimated standard deviation of the mean, i.e., $s_{\bar{x}} = s/\sqrt{N}$. For a sample of size 10, the t value is 2.23, compared with the value 1.96 for confidence limits using the Normal distribution.

A common use of the t-test, is to test whether or not there is a difference between two sample means. The null hypothesis is that both samples (or size n_1 and n_2) are randomly drawn from a single Normal population. Our best estimate of the variance of that population is therefore a *pooled estimate* based on a weighted average calculated from the two sample variances s_1^2 and s_2^2:

$$s_p^2 = \frac{(n_1 - 1)s_1^2 + (n_2 - 1)s_2^2}{n_1 + n_2 - 2} \tag{2.12}$$

On the null hypothesis, the standard error of estimate of the mean is given by

$$s_e = s_p\sqrt{1/n_1 + 1/n_2} \tag{2.13}$$

Note that since two means are calculated, the degrees of freedom of this estimate is $(n_1 + n_2 - 2)$, and this is also the degrees of freedom for the calculated value of t

$$t_c = \frac{\bar{x}_1 - \bar{x}_2}{s_e} \tag{2.14}$$

The following is a MATLAB function that implements this type of t-test:

```
function [m,s] = ttest(x1,x2)
% function [m,s] = ttest(x1,x2)
% x1 and x2 are data vectors
% the function returns their means m(1) and m(2)
% and standard deviations s(1) and s(2)
% and give the result of a t-test for the
% difference between the means, using
% an alpha level of 0.05
n1 = length(x1);            % size of sample 1
n2 = length(x2);            % size of sample 2
m(1) = mean(x1);
m(2) = mean(x2);
s(1) = std(x1);
s(2) = std(x2);
% pooled standard deviation
df = (n1+n2-2);             % pooled degrees of freedom
if df > 30
    error('Degrees of freedom > 30')
    end;
sp = sqrt(((n1-1)*s(1)^2 + (n2-1)*s(2)^2)/df);
se = sp*sqrt(1/n1 + 1/n2);  % standard error of mean
tc = (m(1) - m(2))/se       % computed t
```

```
t5 = [12.7;4.30;3.18;2.78;2.57;2.45;2.37;2.31;2.26;2.23;...
      2.20;2.18;2.16;2.15;2.13;2.12;2.11;2.10;2.09;2.09;...
      2.08;2.07;2.07;2.06;2.06;2.06;2.05;2.05;2.05;2.04];
t = t5(df)                      % theoretical t on null hypothesis
dif = t5(df) - abs(tc);
if dif < 0
    fprintf('\nNull Hypothesis rejected\n');
else
    fprintf('\nNull Hypothesis accepted\n')
end
```

Note that this function uses a "look-up table" instead of a mathematical function to calculate the theoretical value of t. Functions that calculate good approximations to many statistical distributions are available, both from the The MathWorks, Inc. and elsewhere (see end of this chapter).

Data suitable for use with `ttest` is provided by the files `lit1.dat` and `lit2.dat`. They provide analyses for silica and titania for samples of low and medium grade schists in the Littleton Formation of New Hampshire (see Chapter 4 for further discussion). To apply a t-test for the difference between the silica means, load these two files from the disk, and use only the first columns as the two vectors:

```
[m,s] = ttest(lit1(:,1),lit2(:,1))
```

The result is that the calculated t_c is -0.8152, the tabulated theoretical t value is 2.07 (so sample values are expected to lie between -2.07 and $+2.07$), and therefore the null hypothesis is accepted, i.e., there is no difference between the means of the two samples, at the $\alpha = 0.05$ level.

In the preceding discussion, we have assumed that there are two ways in which we might reject a null hypothesis about the mean: the sample mean might be too low, or it might be too high for us to accept. This type of test is called a "two tailed" test. Setting confidence limits always assumes that both types of variation are possible, but null hypotheses are not always of that type. For example, suppose we analyze for pollutant x in samples of groundwater. Any level greater than μ_0 constitutes a danger to health. We are interested only in the null hypothesis that $\mu \leq \mu_0$, so the confidence level for the test refers only to our confidence that the mean content of the pollutant is *less than* μ_0. This is called a "one tailed" test. Unfortunately, published tables of the t-distribution either give α levels for two-tailed tests, or for one-tailed tests. Check carefully which is the case for the tables you use. Then if the table is for a one-tailed test (e.g., the table in Davis, 1986, p.62), you must halve the α level to use the table for a two-tailed test, or for setting confidence limits.

For testing hypotheses about variances, we can use either the Chi Square test, or the F-test, depending on the type of hypothesis. If we want to compare a sample variance s^2 with a hypothetical population variance σ^2, we calculate

$$\chi^2 = \frac{(N-1)s^2}{\sigma^2}$$

and compare it with the tabulated values for $(N-1)$ degrees of freedom. For a two-sided test, we can reject the null hypothesis at the 95% level if either the observed value is *less than* the tabulated value for $\alpha = 0.975$, or if it is *greater than* the tabulated value for $\alpha = 0.025$. Note that the Chi Square distribution is not symmetrical, like the Normal and t-distributions, so it is necessary to look up two different values. Similarly, 95% confidence intervals for the variance can be set up using

$$(N-1)s^2/\chi^2_{0.025} < s^2 < (N-1)s^2/\chi^2_{0.975}$$

Another important hypothesis about variances that we may wish to test is the following: suppose we have two independent estimates of variances s_1^2, s_2^2, where $s_1 > s_2$. Each estimate has its own degrees of freedom: if the estimates are derived from two different samples of sizes N_1, N_2, the degrees of freedom are $(N_1 - 1)$ and $(N_2 - 1)$ respectively. We want to test the null hypothesis that both variances are estimates of the variance of a single Normal population (which is equivalent to saying that there is no 'significant' difference between the two estimates of the variance). We can do this by computing the statistic $F = s_1^2/s_2^2$ and comparing with the tabulated value of F for the appropriate α level, and two degrees of freedom. For example, if the variances are 40 and 20, based on sample sizes of 13 and 17, we compare $F = 2$ with the tabulated value for $\alpha = 0.05$ and 12 and 16 degrees of freedom. That value is 2.42, so we accept the null hypothesis at the 95% confidence level.

In the chapters that follow, we will present other applications of the sampling distributions that we have described here.

2.8   Types of Errors

The confidence level, e.g., 95% confidence, has as its complement a probability that the null hypothesis is rejected when it is actually true. This is called the probability of a Type I error, α. For a 95% confidence level it is 0.05. Its level can be selected by the investigator to match the situation. For example, if an error of this type could lead to a consumer falling ill because of a high level of water or air pollution, then the level should be set lower than the standard 5 or 1%.

Another type of error takes place when a null hypothesis is accepted when it is actually false. This is called a Type II error, and its probability is designated as β. Unfortunately, its level generally cannot be predicted. The reason it cannot be predicted is that, in order to do so, we have to know exactly what the true alternative is to the null hypothesis. It cannot even be controlled, except by increasing the probability of a Type I error, or by increasing the sample size.

There are examples where a Type II error could have more serious consequences than a Type I error. In scientific research, the investigator generally hopes to be able to disprove the null hypothesis. For example, a petrologist hopes to be able to show that continental basalts have less titanium (or some other element) than oceanic basalts. If they do, then the titanium content might be used to identify ancient basalts whose original source is in doubt. The null hypothesis is that both basalts have equal amounts of titanium. The initial investigation might be based on analysis of just a few samples from each group. If a t-test for identity of the population means is accepted, the consequence might be that this line of research is abandoned. The small sample size means, however, that there is a large probability that a small real difference in titanium between the two populations might be missed. By lowering the confidence level for tests performed during a preliminary investigation, the petrologist might conclude that at least there was a sufficient probability that a difference exists to encourage further investigation. When many more data have been collected, a more rigorous test, using a lower probability of Type I error, would be appropriate.

2.9 Software

The MathWorks, Inc. sells a Statistics Toolbox that implements many statistical routines. A less extensive, freeware statistics toolbox, called `Stixbox` has been developed by Anders Holtsberg and is available from `ftp.maths.1th.se`. It includes functions that generate Chi Square, Student's t, and F distributions (and many others). The ready availability of such functions is one reason why no statistical tables are included in this book.

Many commercial software packages have been developed for statistical work. They are all different, implemented on different computers, and generally not only provide many options, but are also expensive. Professional statisticians have developed their own computer language `S plus`. It too, is not cheap, but at least it provides a common language, well adapted to statistical programming. It is described by Becker et al. (1988) and Spector (1994), and is available from StatSci, 1700 Westlake Ave. N, Suite 500, Seattle, Washington 98109. `statlib` is also the name of a computer archive of statistical routines, written in several different languages. It can be reached by ftp at `lib.stat.cmu.edu` (login with the username `statlib`).

2.10 Recommended Reading

There are many introductory texts on statistics, including several on applications in the earth sciences. Perhaps the best of these are

Davis, John C., 1986, Statistics and Data Analysis in Geology. New York, John Wiley and Sons, second edition, 646 p. (QE48.8.D38)

Berry, D.A. and B.W. Lindgren, 1996, Statistics: Theory and Methods. Belmont CA, Duxbury, second edition, 702 p. (QA276.12.B48 A modern text on classical statistics, including more discussion than usual of probability theory and the Bayesian approach.)

The following are (relatively) inexpensive paperbacks:

Evans, M. and N. Hastings, 1993, Statistical Distributions. New York, Wiley Interscience, second edition, 170 p. (QA273.6.E92 More advanced treatment of the distributions discussed in this chapter, plus several others.)

Lowry, Richard, 1989, The Architecture of Chance: an Introduction to the Logic and Arithmetic of Probability. Oxford University Press, 175 p. (QA273.L685 A more extended discussion of probability than can be given in this book, but still at an elementary level.)

Norman, G.R. and D.L. Streiner, 1986, PDQ Statistics. Toronto, B.C. Decker, 172 p. (QA276.12.N67 Irreverant, compressed style: covers a lot of territory.)

Silver, S.D., 1970, Statistical Inference. London, Chapman and Hall, 191 p. (QA276.S5 A serious work on the foundations of this subject.)

The following is a well-written review article, written from the modern perspective of nonlinear dynamics.

Eubank, Stephen and Doyne Farmer, 1990, An introduction to chaos and randomness *in* Jen, Erica, ed., 1989 Lectures in Complex Systems. Reading, MA, Addison-Wesley Publ. Co., Inc., Santa Fe Institute Studies in the Sciences of Complexity, Lectures v.II, p.75–190 (QA267.7.L44).

Chapter 3

Algorithms and Computing

3.1 Introduction

In the last two Chapters, we have seen examples of how easy it is to compute some simple statistics and graphic displays using MATLAB. This is because MATLAB is a "high level" computer language: we do not have to program all the steps of a computation, but can use commands such as `mean`, `std` and `erf` to calculate complex functions, because the computational steps have already been programmed by the people who wrote MATLAB.

Most scientists do not want to become computer programmers, and they particularly do not want to have to write the tedious routines that are necessary to input and output data from and to files, or to display and print good quality graphics. The fact that MATLAB does these tasks for you is one of its main attractions. Sooner or later, however, even a user of MATLAB has to become more concerned about the nuts and bolts of computing. The purpose of this chapter is to review the minimum knowledge about computer programming necessary to use MATLAB (and other high level computer languages) efficiently and intelligently.

A computer consists of a combination of a *central processing unit (CPU)*, which performs elementary operations such as adding binary numbers, several different kinds of *memory*, and several different kinds of *input and output (I/O)* devices. A computer performs useful tasks by following a *program*, that is a sequence of instructions, stored in the memory. All of the basic operations are carried out using binary arithmetic, but luckily the programmer is rarely called upon to operate at that level. Even assembly language (ASM) is well removed from the basic machine language, which is determined by the particular CPU used. The lowest level language in common use in the 1990s is the *C language*: other common languages at only slightly higher levels are *FORTRAN, Pascal*, and *BASIC*.

There are two different types of computer-program processors:

- **interpreters** that interpret each command (i.e., convert it into instructions that can be carried out by the CPU) and then carry out those instructions before continuing to the next command;

- **compilers** that first compile the entire program (the "source code") into CPU instructions, and store it in memory (generally as a program, called the "compiled" or
  "executable code"), and only then carry out the complete set of instructions.

Any computer language may be processed either by an interpreter or by a compiler. The form
of BASIC commonly supplied with MS-DOS or WINDOWS (QBASIC) is an interpreter,
and there are even C interpreters (written for instructional purposes), but implementations
of most low-level languages are compilers. The reason is that compiled programs run much
faster (one or two orders of magnitude faster!) than interpreted programs—though first
they have to be compiled and "de-bugged" so that they perform correctly, and that may
take some time. MATLAB is an interpreter: each command is performed as it is typed in.
MATLAB programs (called *scripts* or user-defined *functions*) are also performed command
by command. MATLAB can still perform at high speeds, because many complex operations
(such as matrix operations and functions such as `erf`) are precompiled, but to make best
use of MATLAB it is necessary to write scripts in a form that makes best use of MATLAB's
strengths and avoids its weaknesses (see Appendix A).

3.2   Algorithms

An algorithm is a precise description of how to solve a problem, using a well-defined sequence of instructions. A recipe is an algorithm (Shore, 1985), and so are the directions
for finding a friend's house—though in both cases the precision may leave something to be
desired. Unlike people, computers can only be given instructions that are absolutely precise and unambiguous—they never use "common sense" to interpret programs. Computer
languages are designed to provide just such a set of instructions.

All numerical problems can be expressed as an algorithm. In simple cases, ordinary
languages can express the algorithm. For example, to compute the arithmetical mean of a set
of N real numbers, add them up and divide the sum by N. We express this in mathematical
notation by the equation

$$\bar{x} = \sum_{i=1}^{N} x_i$$

The following are the ways that this algorithm is expressed in various computer languages:

- BASIC

```
SUM = 0
FOR I = 1 TO N
    SUM = SUM + X(I)
NEXT I
XBAR = SUM/N
```

- C

```
sum = 0;
for (i = 1; i = n; i++) sum = sum + x(i);
xbar = sum/n;
```

- FORTRAN

```
      SUM = 0
      DO 10 I = 1, N
         SUM = SUM + X(I)
   10 CONTINUE
      XBAR = SUM/N
```

- Pascal

```
Sum := 0.0;
For i = 1 to N do Sum := Sum + x[i];
Xbar := Sum/N;
```

Take a minute to read through these examples carefully: which do you think is the easiest to understand, for those who know little or nothing about programming? Probably most people would rank them in order of increasing obscurity: BASIC, Pascal, FORTRAN, C. Each program performs the calculation in the same way: first we set a variable Sum to zero, then use a for (or do) loop to add all N values in the array x(i) to Sum, then determine the mean Xbar by dividing Sum by N. The languages differ only in relatively minor details. For example Pascal and C distinguish between *assignment* (denoted := in Pascal, and = in C) and *identity* (denoted = in Pascal and == in C). This is really quite an important distinction: $S = S + x_i$ makes no sense in ordinary mathematical notation, which uses the equals sign exclusively for identity. Computer languages use the equal sign, or some variant of it, to mean assignment, that is, replacing the value of the symbol on the left hand side of the equation, with the computed value of the quantities on the right hand side. MATLAB, like C, uses the equals sign (=) to mean assignment, and the double equals sign (==) to mean identity.

In this book, no attempt is made to teach any of the common low level languages, let alone their high level "object oriented" extensions (e.g., C++). It is useful, however, to have reference to a language that is easy to read, so that it can be used as a way of writing down algorithms. Many texts use "pseudo-code:" a set of instructions that are almost, but not quite, written in a computer language. Most pseudo-codes are based on Pascal, a language originally written for instructional purposes. The source code for MATLAB, however, is written in C, and some elements of the C language have been incorporated into MATLAB.

Finally, note that not all mathematical equations can be considered to be algorithms. An equation like Equation 2.8 (in Chapter 2, p.21) defines a mathematical function (in this example, the cumulative Normal distribution) without indicating how it can actually be evaluated. Often several different numerical algorithms can be used to evaluate a single mathematical function. Which is the best method to use depends on many factors: the nature of the data, and the speed and precision required. These factors are explored fully by the science of *numerical analysis*: its scope is too large to be covered by this book. For the most part, MATLAB relieves the user from decisions about numerical methods: MATLAB functions use numerical algorithms originally developed in the public domain LINPACK and EISPACK projects. To explore this subject further, see Lindfield and Penny (1995).

3.3   Structure of a Computer Program

Computer programs, even scripts written in MATLAB, have a definite structure. It varies somewhat from language to language, but roughly follows the following order (typical of Pascal):

```
Program Name;
Declarations
   (definition of constants and variable types, etc)
Procedures
   (called subroutines in other languages)
Begin  {Main program---Pascal uses braces for enclosing comments}
   instructions
End. {Main program}
```

Some languages require that the type of each constant or variable be defined before the variable is used. Examples of different "types" are integers, real numbers, complex numbers, and "boolean" (true/false) variables. In MATLAB all variables are assumed to be matrices (arrays of numbers): vectors and scalars are simply special types of matrices. No formal distinction is made between integers and real numbers. No special type declarations are necessary.

Constants and variables defined by the main program are called *global*. Those defined by procedures (subroutines) are called *local* because they have meaning only within the subroutine, unless their definition and value is passed to the main program (as one of the "output arguments" of the subroutine). In MATLAB the main program is called a *script*, and procedures or subroutines are called *functions*: both are saved in memory as *M-files*. The distinction is important because the values of variables defined within functions are lost when control is transferred back to the main program, unless they are passed as output arguments. The advantage of using functions is that they are self contained units that can be tested separately, and that the same symbols can be used in different functions for different quantities. What is x in one function has nothing to do with what is x in another function.

In general, a well written program should have only a short main program, whose main purpose is to call various subroutines. The subroutines may be defined as part of the program, or they may be subroutines written independently, and included in the program by the declarations. In MATLAB, these two types correspond, roughly, to functions defined by the user, or functions defined by MATLAB or one of its Toolboxes.

Though a function may be defined using one set of symbols for arguments, it is generally called using different symbols. We have already seen examples of this. The MATLAB manual defines sort using the arguments Y and X: Y = sort(X). In the script given in Chapter 1 (p.9), however, sort was called using the output argument x and the input argument dewijs, i.e., by entering x = sort(dewijs).

Note, incidentally, that MATLAB distinguishes between upper and lower case letters. A habit worth cultivating in writing MATLAB scripts and function is to use lower case letters for scalars or vectors, and upper case letters for matrices. Like all computer programs, MATLAB scripts and functions are written using only ASCII symbols (see Appendix I). So the ith element of a vector x_i is represented in MATLAB scripts as x(i).

As an example, of these ideas, consider a rewritten version of normfit, originally presented in Chapter 2.

```
function [f, chi2] = normfit2(y, x, c)
% [f, chi2] = normfit2(y,x,c)
% plots a histogram of the data in the vector y,
% using the vector of class limits given in x.
% The class intervals must all be equal, and the
% range of x should include most of the data.  n is the
% vector of frequencies in each class.  Also plots
% the fitted normal distribution, and calculates ChiSquared
% If c = 1, then the histogram will be color-filled
% written by Gerry Middleton, April 1998.
f = hist2(y,x,c);        % plot histogram
hold on;
m = mean(y);
s = std(y);
chi2 = nchisq(x,f,m,s);   % calculate Chi Square
nplot(x,f,m,s);          % plot Normal curve
title('Histogram and fitted Normal curve');
xlabel('x'), ylabel('frequency');
hold off
```

This version performs exactly the same tasks as the first version, and runs at very nearly the same speed. It has the advantage that the structure of the function is clearer, i.e., first it plots the histogram using hist2, then it calculates Chi Square using nchisq, and finally it plots the fitted Normal curve using nplot. To see the details of how this is done we must examine these three user-written functions (they are not given in the text, but are included with the other M-files distributed with this book). A further advantage of splitting a long function into smaller functions is that each of the smaller functions may be used in other programs.

An important part of any program is played by *control statements* that allow looping or branching of the program. There are three fundamental control statements, implemented in almost all languages:

- **Sequence**. This simply means that one statement is automatically followed by the next. In MATLAB, successive statements are separated by carriage returns, or by a comma, or semicolon. Use of a semicolon suppresses display of the result of the statement.

- **Iteration**. These cause loops to be repeated until some limit is reached. MATLAB provides two variants: `for` loops, which use a counter to control the number of loops, and `while` loops, which carry out the loop until some computed limit is reached or exceeded. Loops may not be executed at all, if the limit (set by some earlier statement) is exceeded at the first test.

- **Selection**. These cause branching of the program. MATLAB provides `if-elseif-else` for multiple branching, and `if-break` for jumping out of a loop.

MATLAB's control statements are relatively sparse, compared with those available in languages such as Pascal and C. This is partly because use of control statements in an interpreted language like MATLAB causes the program to run slowly. MATLAB was designed for fast computation of operations on vectors and matrices. It does not support advanced uses of control structures, such as recursion.

As far as possible, in MATLAB, loops should be replaced by *vectorizing* the code. The following script is an example of what this means:

```
% script sumsqrs
% computes the sum of squares of a vector of random
% numbers, by two different methods, and times each.
x = rand(5000,1);    % column vector of 5000 random numbers
% first method: using a for loop
tic;
s = 0;
for i = 1:5000
    s = s + x(i)*x(i);
end
t1=toc;
% second method: using vector operations
tic
s2 = x'*x;
t2=toc;
% print the two sums, and the ratio of the times
s
s2
t1/t2
```

`t=toc` is a MATLAB command that returns the elapsed time (in seconds) since `tic` was set. Note that the last three commands are printed to the screen, because they are not followed by a semicolon.

On a 486 PC, this script indicates that the vector method is almost 100 times faster than the loop method. The vector method makes use of the fact that the vector product of the transpose of a column vector x' times the vector is the sum of squares of the elements (see next chapter).

Another example is the use of the function cumpinv listed in Chapter 2 (p.24). Examining this function, one might get the impression that it could be used only to return a single z for a single p. In fact this is not the case: it will return a vector of z values, corresponding to a vector of p values. In fact, it is used this way in the function cumprob. Note that using a for loop to apply cumpinv (or any other math function, such as the trigonometric functions supported by MATLAB) to calculate a vector of z values, would be a much slower way of doing things.

If vectorizing is not possible, for loops can be speeded up by a simple trick explained in the Manual: preallocate memory space for the total computed result *before* entering the for loop. For example, to allocate space for a column vector with 50 elements, first create a vector y of this size with zero elements using the command y = zeros(50,1);. For other examples, and MATLAB tricks to speed up programs, see Appendix A.

3.4 Data Structures

The title of a famous book written by Niklaus Wirth (1975) is "Algorithms + Data Structures = Programs." Algorithms and program structure are only part of the computing story: the other part is the way that the data are organized.

Most low-level languages are designed to handle integers and real numbers, but not complex numbers, vectors, and matrices. Functions can be written that do manipulate these structures, but they are not part of the basic language. MATLAB has been written specifically to make programming with complex numbers, vectors, and matrices as easy as programming with scalars.

What MATLAB loses by doing this, is the ability to use *pointers*. Therefore MATLAB cannot readily construct data structures such as *linked lists* and *binary trees*. In these structures, each datum is linked to other data in the list or tree, by "pointers" that indicate the memory locations below, and to the left and right of any location in the tree. This means that the programmer has more or less direct control of how the random access memory (RAM) is assigned to data. Data structures such as lists and trees are particularly useful for fast sorting and retrieving data from computer data banks. With control of memory, however, come responsibilities which may be more than the busy scientist can handle.

If it happens that a programming problem cries out for the use of pointers, this does not mean that MATLAB must be completely abandoned. The alternative is to write a function in FORTRAN or C, and compile it for use as a MATLAB MEX-file. For details of these advanced techniques see the appropriate MATLAB manuals.

3.5   Recommended Reading

Lindfield, G. and J. Penny, 1995, Numerical Methods Using MATLAB. New York, Ellis Horwood, 328 p. (QA297.P45)

Schneider, G.M., S.W. Weingart and D.M. Perlman, 1982, An Introduction to Programming and Problem Solving with Pascal. New York, John Wiley, second edition, 468 p. (QA76.73.P2S36)

Shore, John, 1985, The Sachertorte Algorithm, and Other Antidotes to Computer Anxiety. New York, Penguin, 269 p. (QA76.9.C64S56)

Wirth, Niklaus, 1975, Algorithms + Data Structures = Programs. Englewood Cliffs NJ, Prentice-Hall, 366 p. (QA76.6.W56 An analysis of programming fundamentals by the author of the Pascal language.)

Chapter 4

Data as Matrices

4.1 Introduction

In earlier chapters, we introduced the terms *vector* and *matrix* to refer to one-dimensional and two-dimensional arrays of numbers, and remarked that MATLAB was designed especially to perform computations on these data types. In this chapter, we will review some of the elements of matrix algebra, and show why matrices are a particularly useful way of representing most earth-science data.

Matrix algebra arose historically from the need for a concise notation for representing operations on sets of linear equations. For example, consider the two equations:

$$a_{11}x_1 + a_{12}x_2 = b_1 \qquad (4.1)$$

$$a_{21}x_1 + a_{22}x_2 = b_2 \qquad (4.2)$$

These equations relate two variables x_i to two sets of numerical coefficients a_{ij} and b_i. Note that the first subscript of a_{ij} refers to the equation (row) number, and the second subscript refers to the variable that it is associated with. If the variables are arranged column by column, this is also the column number. The equations are *linear* because there are no terms involving powers or cross-products of any of the variables. We know from elementary linear algebra that such a set of equations can easily be solved to yield an expression for the x_i in terms of the coefficients. For such a simple system, the solution for x_1 can readily be found by multiplying the first equation by a_{22} and the second equation by a_{12} and subtracting the second equation from the first. After some rearrangement, this gives

$$x_1 = \frac{a_{22}b_1 - a_{12}b_2}{a_{11}a_{22} - a_{12}a_{21}} \qquad (4.3)$$

The two equations can be represented more compactly using the conventions of matrix algebra. Rewriting the equations as a matrix product gives

$$\begin{bmatrix} a_{11} & a_{12} \\ a_{21} & a_{22} \end{bmatrix} \begin{bmatrix} x_1 \\ x_2 \end{bmatrix} = \begin{bmatrix} b_1 \\ b_2 \end{bmatrix} \qquad (4.4)$$

The arrays of numbers or algebraic symbols are called *matrices* and can be represented using their typical elements by the notation $[a_{ij}]$, $[x_i]$, $[b_i]$ or simply by the bold face letters **A, x, b**. Thus, we may write Equation (4.4) as $[a_{ij}][x_i] = [b_i]$ or

$$\mathbf{Ax = b} \qquad\qquad (4.5)$$

In print, matrices and vectors are generally represented by bold face symbols: capitals for matrices, and lower case for vectors. Vectors can be considered to be just a particular form of matrices: *row vectors* have one row, and several columns, and *column vectors* have several rows, but only one column. A matrix, like **A**, that has the same number of rows and columns is called a *square matrix*.

The algebra of matrices (in particular, matrix multiplication) is defined so that linear equations can be written in this way (see later section).

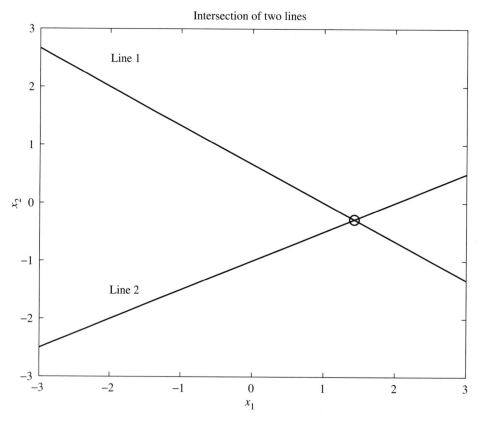

Figure 4.1   Intersection of two straight lines

4.2 Determinants

It is also useful to define a quantity called the *determinant* of a square matrix. For a 2×2 matrix $\mathbf{A}$ this is defined as

$$|\mathbf{A}| = |a_{ij}| = (a_{11}a_{22} - a_{12}a_{21}) \tag{4.6}$$

$$= (a_{22}/a_{21} - a_{12}/a_{11})a_{21}a_{11} \tag{4.7}$$

Using this notation, we can represent the solution of the set of two linear equations by the equations

$$x_1 = \frac{a_{22}b_1 - a_{12}b_2}{|\mathbf{A}|} \tag{4.8}$$

$$x_2 = \frac{a_{11}b_1 - a_{21}b_2}{|\mathbf{A}|} \tag{4.9}$$

Notice that this also tells us that the condition that a solution exists is that $|\mathbf{A}| \neq 0$.

We can see the geometric interpretation by plotting a graph for a particular case, using the following MATLAB script

```
% script intersec.m
% plots the intersection of two lines,
% and demonstrates that the point of intersection
% is the same as the solution of the set of equations.
A = [2 3; -1 2];  % define the A matrix
b = [2; -2];      % define the column vector b
x1 = [-3 +3];     % x1 limits
x2 = (b(1) - A(1,1)*x1)/A(1,2);  % x2 vector for first line
xx2 = (b(2) - A(2,1)*x1)/A(2,2); % x2 vector for second line
disp('intersection coordinates: ');
x = A\b                          % solution of equation system
plot(x1,x2,'r','LineWidth',1);
hold on;
plot(x1,xx2,'y','LineWidth',1);
plot(x(1),x(2),'o');
title('Intersection of two lines');
xlabel('x1'), ylabel('x2');
h1 = text(-2,+2.5, 'Line 1');
set(h1,'Color','r');
h2 = text(-2,-1.5, 'Line 2');
set(h2,'Color','y');
```

First a few comments about the script. Note that x2 and xx2 are vectors, with two rows giving the two values of x_2 corresponding to the two values of x_1 specified in the vector x1. The backslash used in the command x = A\b is MATLAB's way of solving a set of linear equations. In this simple case, the solution could easily be programmed, but for larger sets of equations that would be a major undertaking. In plotting the graphs, we first define the limits of x_1 for the plot, then calculate the corresponding values of x_2. We specify the color ('r' indicates red, 'y' indicates yellow, and the line width ('LineWidth', 1 indicates a somewhat thicker line than used for the axes: the "factory setting" is 0.5 points). Since the plot command is used more than once to construct the same graph, we have to use hold on to prevent the first plot from simply replacing the second. The positioning of labels, using text has to be determined by trial and error. Alternatively, we could have used legend (see Appendix A).

The graph shows that the solution to the two equations is the value of the vector $\mathbf{x} = [x_1 x_2]$ where the two lines intersect. This is the point shown circled on the figure (the script also returns the numerical coordinates 1.4286, -0.2857). The graph also shows that, if the slopes of the two lines were equal, there would be no intersection, hence no solution to the system of equations. The slope of the first line is a_{12}/a_{11} and of the second is a_{22}/a_{21} and equality of these slopes is indeed the condition that is obtained by setting the determinant of $\mathbf{A}$, (Equation 4.7), equal to zero:

$$a_{22}/a_{21} - a_{12}/a_{11} = 0 \qquad (4.10)$$

Determinants can be defined for square matrices larger than 2×2 rows and columns. The definition is given in any text on linear algebra. Determinants of any size are evaluated by MATLAB using the function det(A).

4.3   Matrix Algebra

Matrices have their own algebra. It differs somewhat from the familiar algebra of ordinary numbers (scalars). Before describing this algebra, we note a few terms describing matrices. Matrices are *square* if they have the same number of rows and columns. The *unit matrix* $\mathbf{I}$ is a square matrix, with elements $I_{ij} = 1$ if $i = j$, and $I_{ij} = 0$ otherwise. It can be made using the MATLAB command eye(n) where n is the *order* of the matrix, i.e., the number of rows and columns. The diagonal with the ones in it, sloping from top left to bottom right, is called the *principal diagonal*. MATLAB has a command diag that is useful for manipulating diagonals in matrices (see the Manual for details). MATLAB can make special matrices: e.g., full of zeros zeros(m,n) or full of ones ones(m,n). In MATLAB the order of any matrix can be returned using the command [m,n] = size(A).

- **addition and subtraction**. Two matrices are added or subtracted by adding or subtracting their corresponding elements, i.e., $\mathbf{A} \pm \mathbf{B} = \mathbf{C}$ is true if and only if (iff) $a_{ij} \pm b_{ij} = c_{ij}$. The MATLAB command is C = A + B.

- **transposition.** $\mathbf{B} = \mathbf{A}^{\mathrm{T}}$ is the transpose of $\mathbf{A}$ iff $b_{ji} = a_{ij}$, or in other words the rows of the original matrix become the columns of the transposed matrix. Transposition of a row vector converts it into a column vector. The MATLAB command is B = A'. Square matrices that remain the same under transposition are called *symmetrical matrices*.

- **multiplication by a scalar** is straightforward. $\mathbf{B} = c\mathbf{A}$ iff $b_{ij} = ca_{ij}$, i.e., every element of $\mathbf{A}$ is multiplied by c. The MATLAB command is B = c*A.

- **matrix multiplication.** $\mathbf{C} = \mathbf{AB}$ iff $c_{ij} = \sum_k a_{ik}b_{kj}$. From this definition it is apparent that matrix multiplication is a complicated process, involving summing a large number of scalar products to calculate each element of the matrix product. It is only defined if the number of columns of the first matrix $\mathbf{A}$ is equal to the number of rows of the second matrix $\mathbf{B}$: the product has the same number of rows as $\mathbf{A}$ and the same number of columns as $\mathbf{B}$. Clearly, the *order* of multiplication matters: $\mathbf{AB} \neq \mathbf{BA}$ unless both matrices are square and symmetrical. The MATLAB command is C = A*B.

- **matrix inversion.** Division by a matrix is not defined: but the equivalent operation is defined by multiplication by the inverse of a matrix. The system of linear equations

$$\mathbf{Ax} = \mathbf{b}$$

is solved, formally, by premultiplying each side by the inverse $\mathbf{A}^{-1}$ of $\mathbf{A}$:

$$\mathbf{x} = \mathbf{A}^{-1}\mathbf{Ax} = \mathbf{A}^{-1}\mathbf{b}$$

The inverse is defined as the matrix (if it exists) whose product with the original matrix yields the *unit matrix* $\mathbf{I}$

$$\mathbf{A}^{-1}\mathbf{A} = \mathbf{I}$$

A unit matrix (of appropriate size) pre- or postmultiplied by a matrix $\mathbf{A}$ leaves the matrix unchanged, i.e., $\mathbf{IA} = \mathbf{AI} = \mathbf{A}$.

Only a square matrix can have an inverse (which is also square), and only square matrices whose determinant is not equal to zero have an inverse. In MATLAB, the inverse of a matrix can be determined using the function inv. Though a system of linear equations may be formally solved by using the inverse matrix (as shown above), it takes more calculation to determine the inverse than to solve the equations. In MATLAB, systems of linear equations are better solved using the backslash function: $\mathbf{x} = \mathbf{A}\backslash\mathbf{b}$. For more details on the backslash operator see the Manual and Appendix A.

Several other matrix operations are defined in matrix algebra, but these will do for the moment. As an example of their use, we consider how matrix operations may be used to calculate the variance of a data vector, i.e., a random sample of a variable x_i $i = 1, 2, \ldots, N$.

SiO$_2$	TiO$_2$
66.6	0.88
58.1	0.65
64.3	1.06
62.9	0.99
59.0	1.00
67.3	0.97
61.4	0.60
62.7	0.94

Table 4.1   Silica and titania percent in low grade schists of the Littleton Formation (from Moss et al., 1995)

4.4 Application to Variance, Covariance and Correlation

For simplicity suppose $N = 3$, and the data have been saved as a column vector (in MATLAB $x = [x(1); x(2); x(3)]$). Then premultiplying this vector by its transpose yields

$$[x_1 x_2 x_3] \begin{bmatrix} x_1 \\ x_2 \\ x_3 \end{bmatrix} = x_1^2 + x_2^2 + x_3^3 \tag{4.11}$$

which is simply the sum of squares of the x_i. Now if we had first determined the mean, and subtracted it from the x_i, the vector product would yield the sums of squares of deviations from the mean—and dividing this by the degrees of freedom $(N - 1)$ gives the variance.

Matrix algebra has an alternative way of manipulating vectors: the *dot product* or *scalar product*:

$$\mathbf{x.y} = x_1 y_1 + x_2 y_2 + x_3 y_3 \tag{4.12}$$

If $\mathbf{y} = \mathbf{x}$ then this operation produced the sum of squares of $\mathbf{x}$. The dot product of matrix algebra should not be confused with the *array operations* of MATLAB. MATLAB's symbol .* denotes array multiplication, which means element-by-element multiplication. So x .* x produces the *vector* $[x(1)^2; x(2)^2; x(3)^2]$. To obtain the sums of squares using array multiplication we would have to write sum(x .* x).

Suppose we have data vectors for two different variables **x1** and **x2**. We could combine these in a single matrix **X**, whose rows were sample numbers, and whose two columns represent the different variables. For example, consider the set of analytical data taken from Moss et al. (1995), shown in Table 4.1. It shows the percent of SiO$_2$ (silica) and TiO$_2$ (titania) in 8 samples of low-grade schist from the Littleton Formation in New Hampshire. The data constitute a matrix with 8 rows and 2 columns, and are stored as the ASCII file lit1.dat. We could calculate the deviations from the mean, and then the *matrix of sums of square and products* (of deviation from the mean), by using the following MATLAB function:

```
function S = sprod(X)
% S = sprod(X)
% calculates the sums of squares and products of deviations
% from the mean of the column vectors forming the matrix X
% Each column of X is a different variable; each row is a
% different sample.
[r c] = size(X); % r, c are the no. of rows and columns of X
Ir = ones(r,1);  % a col vector of r ones
m = mean(X);     % a row vector of means of cols. of X
M = Ir*m;        % a matrix with the means repeated in r rows
XD = X - M;      % deviations from the mean
S = XD'*XD       % sums of square and products matrix
plot(X(:,1), X(:,2), '+')
```

In this function, note the use of ones to generate a column of ones, with the same number of rows as X. This is then used to generate a matrix of mean values, which may be subtracted from X to give deviations from the mean for all variables (columns).

The principal diagonal of the matrix S contains the sums of squares of the variables $S_{jj} = \sum_i (x_{ij} - \bar{x}_j)^2$. The off-diagonal elements are the sum of products between two different variables, e.g., the jth and kth variables $S_{jk} = \sum_i (x_{ij} - \bar{x}_j)(x_{ik} - \bar{x}_k)$. In the application to silica and titania analyses, the numerical values are

$$\mathbf{S} = \begin{bmatrix} 75.45 & 1.70 \\ 1.70 & 0.20 \end{bmatrix} \tag{4.13}$$

Note that "the sums of squares and products of deviations from the mean" is often shortened to "the sums of squares and products," and abbreviated SSP.

The variances and *covariance* of the two variables can be obtained by dividing by the degrees of freedom $N - 1 = 7$:

$$\mathbf{V} = \begin{bmatrix} 10.778 & 0.243 \\ 0.243 & 0.029 \end{bmatrix} \tag{4.14}$$

Notice that the sums of products and covariance matrices are square and symmetrical. This is because the covariance of the jth with the kth variable is the same (by definition) as the covariance of the kth with the jth variable. The covariance is a measure of the joint variability of two variables, just as the variance is a measure of the variability of a single variable. It is hard to grasp the significance of "raw" covariances such as 0.243: the meaning becomes much clearer if the covariance is standardized by dividing by the square root of the product of the two variances. This yields the *correlation coefficient r*. In our example

$$\begin{aligned} r &= \frac{0.243}{\sqrt{10.778 \times 0.029}} \\ &= \frac{1.70}{\sqrt{75.45 \times 0.20}} \\ &= 0.435 \end{aligned}$$

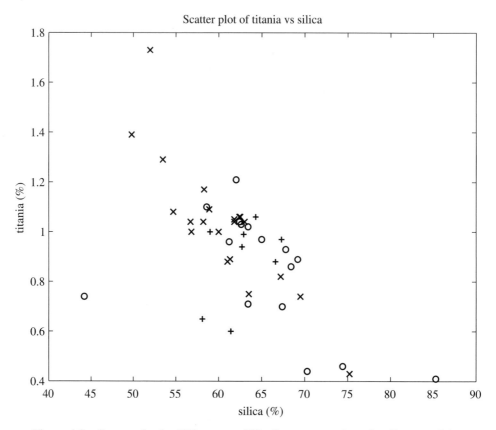

Figure 4.2   Scatter plot for TiO_2 versus SiO_2 for metamorphosed sediments of the Littleton Formation. Crosses (+) indicate low-grade rocks, circles medium-grade rocks, and x's (×) indicate high-grade rocks. Data from Moss et al. (1995).

The correlation coefficient is a number between -1.0 and $+1.0$ which measures the degree of association between two variables. We will see in the next chapter that the square of the correlation coefficient can be interpreted as the proportion of the variance of one variable that can be explained by its association with the other. In our example, $r^2 = 0.189$ so 18.9% of the variance of titania can be explained by its association with silica. The degree of association can also be seen by plotting titania against silica: examining the plot produced by the function sprod shows a large amount of scatter, with only a hint of positive association (see the + symbols in Figure 4.2). Because of the small sample size, we would be inclined to doubt that there was any "real" positive association between the two variables. This is confirmed by looking at a larger sample, including also medium- and high-grade rocks (in lit2.dat and lit3.dat: see Figure 4.2): if anything, it seems that there is a weak negative association.

We have worked through this example, using elementary matrix algebra. If we simply want results, MATLAB provides functions that make this unnecessary. cov calculates the variance-covariance matrix from a data matrix, and corrcoef calculates the corresponding matrix of correlation coefficients.

The following function, however, calculates some further statistics for a data vector x:

```
function [m,m2,m3,m4] = moments(x)
% [m,m2,m3,m4] = moments(x)
% Calculates mean m and 2nd--4th moments
% of a data vector x
N = length(x);
m = sum(x)/N;
m1 = x - m;             % vector of deviations from mean
m2 = (m1'*m1)/(N-1);    % second moment = variance
mm = m1 .* m1;
m3 = (mm'*m1)/(N-1);    % third moment
m4 = (mm'*mm)/(N-1);    % fourth moment
```

The four *moments of a distribution* can be used to describe the shape of distributions: they were used by Karl Pearson to classify frequency distributions, and have been used in geology to describe size distributions (for further discussion, see Miller and Kahn, 1962, Chapter 3). The first moment (about zero) is the mean, the second moment (about the mean) is the variance. The third moment about the mean measures the asymmetry of the distribution. It is used to calculate the *skewness*, generally calculated as $Sk = m_3/m_2^{3/2}$. For symmetrical distributions it is zero, but for asymmetrical ones it can be negative (for distributions with a long tail of smaller values) or positive (for a long tail of larger values). Many natural distributions, like the De Wijs data, are positively skewed. The fourth moment measures how "humped" or "peaked" the central part of the distribution is, and is used to calculate the *kurtosis*, defined as $K = m_4/m_2^2$. For the Normal distribution $K = 3.0$. It is not used much as a descriptive statistic any more because it is subject to large sampling errors.

Try using moments to calculate the skewness and kurtosis of the De Wijs data. Then generate several 100-item samples using x = rand(100,1) and x = randn(100,1) and calculate their sample skewness and kurtosis. A uniform distribution should be very "humped," so have a low kurtosis. Note the large deviation from the theoretical value of 3.0 for the kurtosis of most of the samples generated from the Normal distribution. Figure 4.3 shows the result of such an experiment, using the script skknorm (not printed here, but included on the diskette). It also shows how the error of estimation of skewness and kurtosis decreases with increasing sample size.

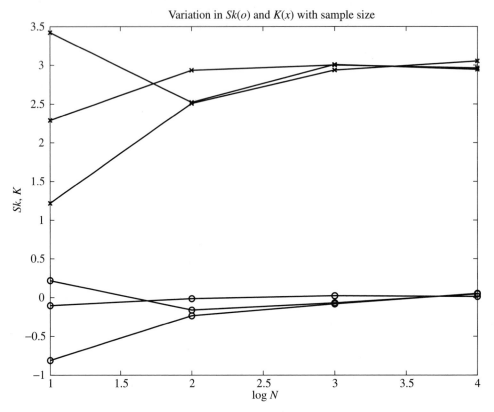

Figure 4.3   Three independent estimates of Sk and K for sample sizes of sizes between
10 and 10,000, drawn from a Normal population.

4.5   Recommended Reading

Davis, John C., 1986, Statistics and Data Analysis in Geology. New York, John Wiley
and Sons, second edition, 646 p. (QE48.8.D38 Chapter 2 describes covariance and
the correlation coefficient; Chapter 3 introduces matrix algebra.)

Ferguson, J., 1994, Introduction to Linear Algebra in Geology. London, Chapman and
Hall, 203 p. (Gives many examples of the application of linear equations and matrices
in the earth sciences.)

Chapter 5

Regression and Curve Fitting

5.1 Introduction

So far in this book, we have considered measurements made only on a single variable. Such data are rare in the earth sciences: generally, variables are measured at a particular place ("station" or "locality"), in which case we can associate at least two spatial coordinates with the variable; or a variable is measured repeatedly at regularly spaced times, in which case we can associate a time with the variable. Often several different variables are measured, for example, in the study of chemical changes during metamorphism, referred to in the last chapter (Moss et al., 1995), 45 different major and trace elements were determined in almost every specimen. The proper techniques for studying such data are called "multivariate," but in this chapter we will be content to study data consisting of samples of just two measured variables.

The first step is always to plot the data as a "scatter diagram" similar to that presented in the last Chapter (as Figure 4.2). Examining such a diagram shows whether or not further analysis is necessary, and shows whether the raw data should be analyzed, or some transformation of the data (such as taking logarithms). It may also indicate if it is useful to fit a straight line to data from the two variables plotted, or if some more complicated fitting technique should be used.

Often, one of these variables can be considered to be the *independent* variable, whose value is fixed (and determined almost without error) before measurement of the other *dependent* variable is possible. For example, many geophysical time series consist of a series of measurements of temperature, or pressure, or speed, or some other quantity, measured at predetermined time intervals (e.g., every hour, or day, or month). Other measurements are made at stations spaced along some traverse: for example, at stations spaced at intervals (not necessarily regular) along a river. In these examples, the times or distances can be fixed rather precisely, so if subsampling (at closely spaced times or places) shows local variability in the dependent variable, it makes sense to think that the variability is inherent in that variable, and not due to the dependent variable.

This is expressed mathematically by using a *regression* model

$$y_i = f(x_i) + \epsilon_i \qquad (5.1)$$

where the y_i are the values of the dependent variable, measured at each of the x_i (predetermined) values of the independent variable. f is some fitted function, such as a straight line or low-order polynomial, and ϵ is a random variable, which is generally considered to consist of unexplained "error" or random natural variation (plus measurement error). For the reasons explained in Chapter 2, such "error" can generally be assumed to be Normally distributed. Note that the very form of Equation 5.1 implies that we believe that all the unexplained "error" resides in y, not in x.

For much of the bivariate data collected in the earth sciences, this belief is not valid. There is often no clear distinction between an independent and a dependent variable. For example, consider the 45 elements measured on samples of the Littleton Formation: which element, if any, is the independent variable? All the analyses are subject to error, and none of the elements can be considered to be the one whose variation "controls" the variation of the others. The same can be said for most geochemical, petrological, and paleontological data. We must be wary of results obtained by applying regression models to such data (see articles by Mann, Troutman, and Williams *in* Size, 1987).

5.2   Linear Regression

In this section we consider in some detail one special case of Equation 5.1: the case where f is a linear function:

$$y_i = b_0 + b_1 x_i + \epsilon_i \qquad (5.2)$$

Note that the fitted line is

$$\hat{y}_i = b_0 + b_1 x_i \qquad (5.3)$$

where $\hat{y}_i$ are the values of y estimated from the fitted line: the ϵ are the differences between the observed values of y and these estimated values. b_1 is the slope of the line, and b_0 is the intercept on the y-axis (the value of $\hat{y}$ when $x = 0$).

As an example, we use some data taken from the classic study of Krumbein (1942). Krumbein measured the mean and maximum size of recent flood gravels collected from the Arroyo Seco, near Los Angeles, CA. The question under investigation was, does the size vary with distance from the source? Some of his data (`asmax2.dat` and `asmean.dat`) are shown in Figure 5.1. Examination of the graph suggests that there does seem to be some variation downstream in maximum size, but not in average size. There is also some irregular local variation, so a linear regression model seems plausible. It is clear from the graph, however, that a straight line will not pass exactly through all the points, for either measure of size. How do we go about obtaining the "best fit" of a straight line to the data?

First we must define what is meant by the "best fit:" it is almost always taken to mean the line that minimizes the sum of squares of the "error" terms in Equation 5.2. This is commonly called the "least squares" method of fitting the line. To apply this method, first

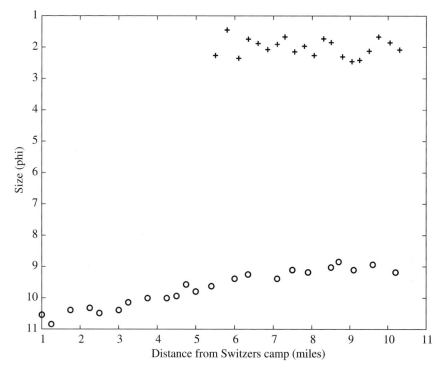

Figure 5.1 Downstream variation in size (in phi units) for gravels in Arroyo Seco (data from Krumbein, 1942). Crosses are mean size, from sieving; circles are maximum size, from 5 largest clasts observed in field.

note the equation for ϵ_i

$$\epsilon_i = y_i - b_0 - b_1 x_i \qquad (5.4)$$

so

$$\epsilon_i^2 = (y_i - b_0 - b_1 x_i)^2 \qquad (5.5)$$

To minimize the sum over i of this quantity we have first to differentiate with respect to the adjustable quantities: they are *not* the variables x_i and y_i (whose values have been fixed by measurement) but the coefficients b_0 and b_1. Using the chain rule of calculus we obtain the following two partial derivatives:

$$\frac{\partial \epsilon_i^2}{\partial b_0} = 2(y_i - b_0 - b_1 x_i)(-1) \qquad (5.6)$$

$$\frac{\partial \epsilon_i^2}{\partial b_1} = 2(y_i - b_0 - b_1 x_i)(-x_i) \qquad (5.7)$$

Then we take the sums over i of these two partial derivatives, and set them equal to zero, which yields the following two equations (remember that summing a constant such as b_0

gives Nb_0):

$$Nb_0 + \sum x_i b_1 = \sum y_i \tag{5.8}$$

$$\sum x_i b_0 + \sum x_i^2 b_1 = \sum x_i y_i \tag{5.9}$$

These are called the "Normal equations," and, as they are a set of two linear equations in the two unknown coefficients b_0 and b_1, they can be solved for b_0 and b_1. Writing the equations in matrix form gives

$$\begin{bmatrix} N & \sum x_i \\ \sum x_i & \sum x_i^2 \end{bmatrix} \begin{bmatrix} b_0 \\ b_1 \end{bmatrix} = \begin{bmatrix} \sum y_i \\ \sum x_i y_i \end{bmatrix} \tag{5.10}$$

The value of the matrix formulation will become more apparent in a later section when we consider multiple regression (regression using more than one independent variable).

In the simpler case considered here, it is easy to derive explicit equations for b_0 and b_1. If Equations 5.8 and 5.9 are written in terms of $x' = x - \bar{x}$ and $y' = y - \bar{y}$, i.e., in terms of deviations from the mean, then they become a single equation

$$\sum (x_i')^2 b_1 = \sum x_i' y_i' \tag{5.11}$$

So if we denote the sums of squares of deviations from the mean (for x) as SS_x, and the sum of products of deviations from the means of x and y as SP, then

$$b_1 = SP/SS_x \tag{5.12}$$

It is also easy to show that the fitted straight line must pass through the mean point ($\bar{x}\ \bar{y}$). Note that if we were to reverse the variables, and calculate the regression of x on y (rather than y on x) the calculated line would still pass through the mean point, but the slope (on the same plot of y vs x) would be different

$$b_1' = SS_y/SP \tag{5.13}$$

The only exception is given by the condition

$$\frac{SP}{SS_x} = \frac{SS_y}{SP} \quad \text{or} \quad r^2 = \frac{SP^2}{SS_x SS_y} = 1 \tag{5.14}$$

In words, the two regression lines are the same only when the square of the correlation coefficient is equal to one: this is the condition that all the points lie exactly along a straight line, with all the error terms (ϵ) equal to zero.

5.3   Application of Analysis of Variance to Regression

There are many ways of applying statistical tests to see if a regression is significant. Some books of mathematical tables include a table of significant r values, for given sample sizes.

source	d.f.	sums of squares	mean squares
Regression	1	$SSR = \sum(\hat{y}_i - \bar{y})^2$	$MSR = SSR/1$
Error	$(N-2)$	$SSE = \sum(y_i - \hat{y}_i)^2$	$MSE = SSE/(N-2)$
Total	$(N-1)$	$SST = \sum(y_i - \bar{y})^2$	

Table 5.1   Analysis of Variance for Linear Regression

The t-test can be applied to test the null hypothesis that $r = 0$, or that $b_1 = 0$ (zero regression slope means that knowing x does not help to predict y, cf. Equation 5.3). But the most useful way of looking at regression, is that it makes it possible for us to divide the total variance of the dependent variable ($s_y^2 = \sum(y_i - \bar{y})^2/(N-1)$) into two independent parts: variance due to regression, and error variance ($\sum \epsilon^2/(N-2)$).

The error term ϵ represents the difference between the predicted and observed value of y, ($\epsilon_i = y_i - \hat{y}$), so it is easy to see that the square of this quantity might be used to estimate the sums of squares (of deviations) about the regression line. But it is not quite so obvious why the degrees of freedom (as defined in Chapter 3) that are needed to convert this sum of squares into an unbiased estimator of the mean is $(N - 2)$. This result is indicated by the following line of argument: In order to calculate, either the total sum of squares or the regression line, it is necessary to calculate the mean—this reduces the degrees of freedom from N to $N - 1$. If the slope is known, as well as the mean, the regression line is fixed—this indicates that the degrees of freedom due to regression is only one. So the error degrees of freedom are what are left: $(N - 2)$.

The calculations are best indicated by means of an Analysis of Variance (ANOVA) table (Table 5.1). Note that the sum of squares column and the degrees of freedom (d.f.) column add up: the mean squares column does not. We have not demonstrated the algebraic identity that the regression sum of squares plus the error sum of squares equals the total sum of squares, but it is done in many statistics texts. Generally we use this identity to reduce the amount of calculation needed, by determining one of these terms by difference.

On the null hypothesis that there is no effect due to regression, the mean squares due to regression, and the mean squares due to error both estimate the same thing—the variance due to error. So the ratio MSR/MSE is expected to follow the F-distribution (see Chapter 2), and the calculated value can be compared with the tabulated value for the chosen α level (say, 0.05) and 1 and $(N - 2)$ degrees of freedom. If it is larger than the tabulated value, the regression is considered significant (the null hypothesis is rejected), otherwise not.

The ANOVA table may be written in a different way (Table 5.2), which is algebraically (and numerically) the same as Table 5.1. This form emphasizes that the square of the correlation coefficient r provides an estimate of the proportion of the total sum of squares that is explained by regression.

MATLAB provides a special command for linear and non-linear regression (`polyfit`). It is used in the function `xyplot` which not only fits a straight line by regression, but also prints out the analysis of variance table:

source	d.f.	sums of squares	mean squares
Regression	1	$SSR = r^2 \sum (y_i - \bar{y})^2$	$MSR = SSR/1$
Error	$(N-2)$	$SSE = (1 - r^2) \sum (y_i - \bar{y})^2$	$MSE = SSE/(N-2)$
Total	$(N-1)$	$SST = \sum (y_i - \bar{y})^2$	

Table 5.2   Alternative Analysis of Variance for Linear Regression

```
function [a, yest, sst, rr] = xyplot(x,y)
% [a, yest, sst, rr] = xyplot(x,y)
% plots y vs x and calculates linear regression, correlation
% and ANOVA
%   x - vector of independent variable
%   y - vector of dependent variable
%   a - vector of fit coefficients: a(1) is slope,
%       a(2) is y-intercept
%   yest - vector of estimated fit points
%   sst  - total sums of square
%   rr - square of correlation coefficient
N = length(x);
Z = [x y];
r = corrcoef(Z); % calculate rr
rr = r(1,2)*r(1,2);
sdy = std(y); % calculate total sums of squares
sst = sdy*sdy*(N-1);
clc;
fprintf('Analysis of Variance Table.\n\n');
fprintf('Source        d.f.      Sums of Squares  Mean Squares\n');
fprintf('----------------------------------------------------\n');
fprintf('Regression    1        %8.2f         %8.2f\n', (rr*sst),...
(rr*sst) );
fprintf('Residuals    %2.0f        %8.2f          %8.2f\n',N-2,...
(1-rr)*sst, (1-rr)*sst/(N-2) );
fprintf('----------------------------------------------------\n');
fprintf('Total        %2.0f        %8.2f\n\n',N-1, sst);
m = 1;
a = polyfit(x,y,m); % calculate regression
yest = polyval(a,x);
fprintf('F = %g\n', (N-2)*rr/(1-rr) );
fprintf('Correlation coeff: r = %g\n',r(1,2) );
fprintf('Equation: y = %g + %g x\n\n',a(2),a(1) );
fprintf('Click on Window to see figure\n');
plot(x,y,'o',x,yest,'-');
xlabel('x'), ylabel('y'), title('Linear Regression');
```

The MATLAB function `corrcoef` is used to calculate the correlation coefficient. This, and the variance of y (obtained by squaring the standard deviation obtained using `std(y)`), are used to calculate the entries in the analysis of variance table. The table is printed out using the `fprintf` command which MATLAB has borrowed from C. For details of the formatting see the Manual. `polyfit`, with the order of the polynomial `m` set to 1, is used to calculate the linear regression coefficients. Once these are known, the estimated values $\hat{y}$ are calculated using `polyval`. Then the points and the regression line are plotted using the `plot` command.

Figure 5.2 shows the results of applying this function to the Arroyo Seco maximum size data. The graph and the analysis of variance are both shown. The calculated F value (229.1) is much larger than the tabulated value, so the regression is highly significant. Close examination of the graph, however, suggests that the line fitted to the data may only be appropriate for the larger maximum sizes (i.e., the most negative Phi values). The conclusion is strengthened by comparing Figure 5.2 with the earlier plot (Figure 5.1). This illustrates the importance of not relying only on computed numbers: whenever possible, plot and examine the actual data! Also note the striking difference in appearance between the two plots, which is due entirely to the difference in vertical scale.

Though MATLAB provides `polyfit` as a function to perform linear regression, it is worth noting that a linear regression can also be performed almost as easily using MATLAB's versatile backslash (or "left division") operator. Suppose that the x and y values are in the vectors x and y, then a linear regression can be performed using the following function

```
function b = lreg(x,y)
% b = lreg(x,y)
% linear regression using the backslash operator
% b is a vector: b(1) = intercept, b(2) = slope
r = length(y);
X = [ones(r,1), x];
b = X \ y;
```

The most common use of the backslash operator is to solve a system of linear equations $\mathbf{Ax} = \mathbf{b}$, where there are n equations, so $\mathbf{x}$ and $\mathbf{b}$ are vectors, and $\mathbf{A}$ is an $n \times n$ square matrix. But it can also be used to obtain a least-squares fit of the data vector $\mathbf{y}$ to a data matrix $\mathbf{X}$, provided the number of data (rows) is the same in $\mathbf{y}$ as in $\mathbf{X}$. This is how the backslash is used in `lreg`. Note that a column of ones is used to obtain the constant term in the regression: if this step was omitted, the regression line would be constrained to pass through the origin.

One advantage of this method is that it is easily modified to allow weighting of the data. As an example of why we might want to do this, consider again the plot shown in Figure 5.2. Because the regression seems to be linear only near the origin, we might want to use the inverse distance to weight the data, i.e., weight each x_i value by $w_i = 1/x_i$. Then the fitted line will fit the points upstream better than it does the points downstream. We do this by generating a diagonal matrix W containing the square roots of the weights in the principal

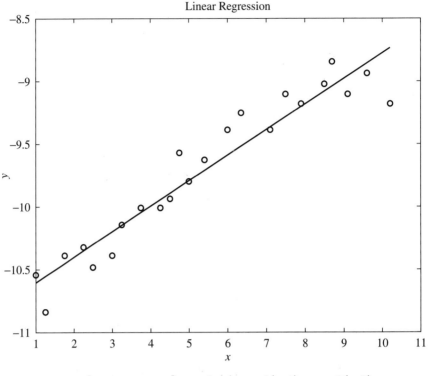

```
>>[a,yest,sst,rr] = xyplot(asmax2(:,1),asmax2(:,2),
Analysis of Variance Table.
Source       d.f.     Sums of Squares      Mean Squares
-----------------------------------------------------------
Regression    1           7.18                7.18
Residuals    21           0.66                0.03
-----------------------------------------------------------
Total        22           7.83

F = 229.069
Correlation coeff: r = 0.957091
Equation: y = -10.8073 + 0.203254 x
```

Figure 5.2   Regression of maximum size vs distance down stream, for Arroyo Seco
gravels.

diagonal using W = diag(sqrt(1 ./ x), and determine the regression coefficients from
b = W*X\W*y.

Another application of Analysis of Variance to regression can be made if there is more
than one value of y measured for at least some of the x values. Then the Residuals Sums
of Squares may be divided into two parts: one part is due to deviations of the data from
linearity, and the other part is an Error term due to deviation of repeated measurements
from the mean y for any x. A second version of xyplot, named xyplot2, implements the

analysis for this case. As an example application, we have included a data file `gal.dat`, with data taken from Shaw and Bankier (1954). The data illustrate a typical calibration problem in analytical geochemistry, and are discussed in detail by Shaw and Bankier in their paper. The data are arranged as a two-column matrix: the first column consists of parts-per-million of a trace element, gallium, in a standard; and the second column is an instrumental reading. The regression line is calculated so that the amount of gallium in unknown samples may be calculated from the instrumental reading. We are also interested in knowing whether the calibration line is linear or not. Application of Analysis of Variance to the raw data shows that it is not; but the calibration becomes linear if we use the logarithm of the gallium values (as Shaw and Bankier did) rather than the raw data.

5.4 Multiple Regression

Multiple regression applies regression techniques to estimating a dependent variable when there are two or more independent variables. When there are two independent variables, we can still visualize this process: x_1 and x_2 are like horizontal spatial coordinates, and y is like a vertical elevation. The estimated values

$$\hat{y}_i = b_0 + b_1 x_{i1} + b_2 x_{i2} \tag{5.15}$$

are points on a plane, making an intercept of b_0 on the vertical axis erected at the x-coordinates origin, and having slopes b_1 and b_2 in the x_1 and x_2 directions, respectively. When there are more than two independent variables, the surface becomes a "hyper-plane:" we can no longer visualize it, but we can still fit it to the data using the same least-squares techniques that we used in the simple one dependent variable case.

In the case of n independent variables, the linear regression model is

$$y_i = b_0 + b_1 x_{i1} + b_2 x_{i2} + \ldots + b_n x_{in} + \epsilon_i \tag{5.16}$$

and applying the least squares criterion lead to the Normal equations

$$
\begin{aligned}
N b_0 + \sum x_1 b_1 + \sum x_2 b_2 + \ldots + \sum x_n b_n &= \sum y \\
\sum x_1 b_0 + \sum x_1^2 b_1 + \sum x_1 x_2 b_2 + \ldots + \sum x_1 x_n b_n &= \sum x_1 y \\
\sum x_2 b_0 + \sum x_1 x_2 b_1 + \sum x_2^2 b_2 + \ldots + \sum x_2 x_n b_n &= \sum x_2 y \\
\ldots \ldots \ldots \ldots \ldots \ldots \qquad & \quad \ldots \\
\sum x_n b_0 + \sum x_1 x_n b_1 + \sum x_2 x_n b_2 + \ldots + \sum x_n^2 b_n &= \sum x_n y
\end{aligned}
$$

To simplify the notation, the subscripts of summation i have been omitted.

As indicated in the last section, this can be represented by a matrix equation (similar to Equation 5.10), which can be written compactly using subscript notation

$$[\Sigma x_j x_k][b_k] = [\Sigma x_k y] \tag{5.17}$$

Here subscripts j and k refer to variables, so $j = 1, 2, \ldots, n$, and $k = 0, 1, 2, \ldots, n$, and we use a "dummy" variable $x_0 = 1$, to obtain b_0, the intercept on the y-axis. The other b_k are the coefficients ("components of slope") associated with each of the independent variables.

We can simplify this equation somewhat by transforming the variables x_{ij} and y_i so that they are deviations from the mean rather than the variables themselves (as we did in the previous section). Then the first equation disappears, because all the summation terms are equal to zero. The set of equations can still be represented by a matrix equation like Equation 5.17, but j and k range from 1 to n and there is no need to make use of a "dummy" x_0. We will use the notation

$$\mathbf{Sb} = \mathbf{p} \tag{5.18}$$

for this form of the Normal equations. $\mathbf{S}$ is the matrix of sums of squares and products of deviations from the mean, $\mathbf{b}$ is a vector of slope-coefficients b_j, and $\mathbf{p}$ is a vector of sums of products of deviations from the mean of the x_j with y.

For a multiple linear regression, the Analysis of Variance table, similar to Table 5.2, is

source	d.f.	sums of squares
due to regression	n	$R^2 t = \mathbf{b}^{\mathbf{T}}\mathbf{p}$
about regression	$N - n - 1$	$(1 - R^2)t$
total	$N - 1$	t

where t is the total sum of squares of deviations from the mean $t = \sum(y - \bar{y})^2$, and R is the multiple correlation coefficient. R is the multidimensional equivalent of the two dimensional correlation coefficient. It is computed from the matrix equation shown in the table ($R^2 = \mathbf{b}^{\mathbf{T}}\mathbf{p}/t$). Note that each independent variable contributes one degree of freedom to the regression. If regression is performed step-by-step, first for one independent variable, then for two, and so on, then the contribution of each new variable can be calculated by difference. This makes it possible to test whether or not adding a new variable adds significantly to the regression.

MATLAB can be used to generate the $\mathbf{S}$ matrix and $\mathbf{p}$ vector and to solve for $\mathbf{b}$ using the backslash operator b = S\p. b_0 can be determined by substitution of the mean values of y and the x_j in Equation 5.15. The analysis of variance table can then be readily calculated. Alternatively, multiple linear regression can also be performed directly using the backslash operator. The function lreg, presented in the previous section, will do this, if x is the matrix of x_{ij} data, but it does not perform analysis of variance.

The following is an example of multiple regression in the earth sciences. An interesting problem in plate tectonics is what causes the movement of the plates: a variety of causative mechanisms have been suggested including the pull of subducting plates sinking into the mantle, or the push of hot volcanic material rising up at the mid-ocean ridges. One way to investigate which are the most important mechanisms is to measure the absolute rate of movement of the plates, and compare those rates with the geometric properties of the plates. This was the approach taken by Forsyth and Uyeda (1975), and Table 5.3 gives data taken from their paper. Measurements were made on 12 major plates, shown in Figure 5.3.

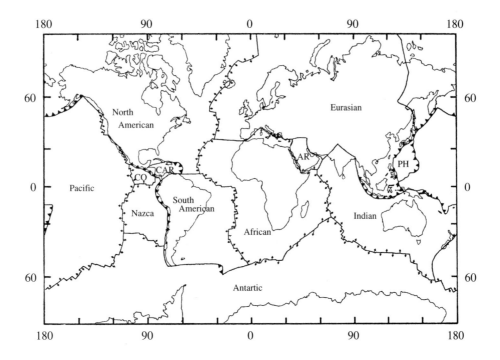

Figure 5.3   Tectonic plates studied by Forsyth and Uyeda (1975).

The dependent variable, plate speed, is the speed relative to the interior of the earth, as far as this can be determined from the apparent movement of hot-spots and other evidence (see Cox and Hart, 1986, for a good introduction to plate tectonics). The independent variables, are total plate area x_1, area of continental part x_2, effective length of ridge x_3, and effective length of trench x_4. Forsyth and Uyeda hypothesized that the speed would be related not so much to the total length of the plate boundaries as to their *effective* lengths, defined as "the length of the boundary which is capable of exerting a net driving or resisting force. For example, two mid-ocean ridges on opposite sides of a plate exert no net force on the plate because their effects cancel." The data are shown in Table 5.3.

To apply multiple linear regression, we must first decide the relative importance of the independent variables, then add each variable in turn until there is no improvement in the fit. The single most important variable may be chosen by determining which one has the highest correlation with the dependent variable. Unfortunately, it is one of the disadvantages of multiple regression that there is no simple way to determine in advance which is the next most important variable: strictly, it has to be determined by trial. One method is to determine the *partial correlation coefficients* of the remaining independent variables with the dependent variable after the effect of the first variable has been removed, and then choose the next highest correlation, and so on. We will use the simpler, but more fallible, method of trying the variables in order of correlation. The matrix of all possible

plate	total area $\times 10^6$ km^2	continental area $\times 10^6$ km^2	effective ridge length $\times 10^2$ km	effective trench length $\times 10^2$ km	speed (cm/yr)
NA	60	36	86	10	1.1
SA	41	20	71	3	1.3
PAC	108	0	119	113	8.0
ANT	59	15	17	0	1.7
IND	60	15	108	83	6.1
AF	79	31	58	9	2.1
EUR	69	51	35	0	0.7
NAZ	15	0	54	52	7.6
COC	2.9	0	29	25	8.6
CAR	3.8	0	0	0	2.4
PHIL	5.4	0	0	30	6.4
ARAB	4.9	4.4	27	0	4.2

Table 5.3   Plate geometry and motion: from Forsyth and Uyeda (1975).

correlations between both dependent and independent variables is easily determined using the MATLAB command corrcoef and is shown in Table 5.4.

The highest (positive) correlation with y is shown by x_4 (effective trench length), with a slightly larger negative correlation shown by x_2 (continental area). Other correlations are rather low, bearing in mind the small sample size (only 12 modern plates).

The first analysis of variance table, for regression of y on x_4, is readily calculated, using the square of the correlation coefficient, ($r^2 = 0.5254$) and the total sums of squares of deviations $t = 97.9767$ (see Table 5.2):

source	d.f.	sums of squares	mean squares	F
Regression	1	$r^2 t = 51.47$	51.47	11.07
Error	10	$(1 - r^2)t = 46.51$	4.65	
Total	11	$t = 97.98$		

The F value in tables for $\alpha = 0.05$ and 1 and 10 degrees of freedom is 4.96—so clearly the null hypothesis must be rejected and there is a significant regression between plate speed and effective trench length. Figure 5.4 shows the graph produced by xyplot.

variable	x_1	x_2	x_3	x_4	y
x_1	1.00	0.52	0.70	0.40	-0.22
x_2	0.52	1.00	0.19	-0.39	-0.74
x_3	0.70	0.19	1.00	0.69	0.17
x_4	0.40	-0.39	0.69	1.00	0.72
y	-0.22	-0.74	0.17	0.72	1.00

Table 5.4   Table of correlation coefficients for plates

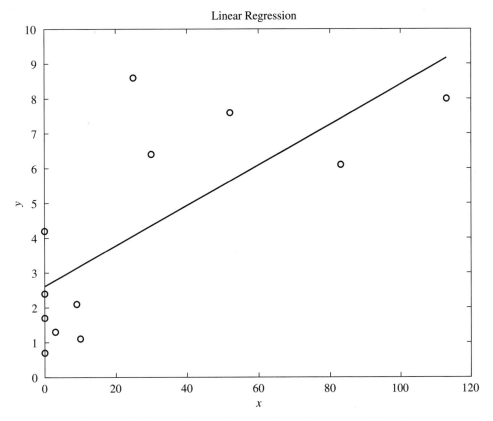

Figure 5.4   Regression of plate speed on effective trench length

To calculate a multiple regression we use the following MATLAB function:

```
function [b, SR, t] = mreg(X)
% [b, SR, t] = mreg(X)
% mreg(X) performs a multiple linear regression on the data
% in the array X: this array has N rows (samples) and p columns:
% the first p-1 columns are measurements on independent
% variables (or squares, products, etc. calculated from
% such data).  The output consists of B, the vector of
% regression coefficients; SR, the sum of squares due to
% regression; and t, the total sum of squares.
[N,p] = size(X);
m = mean(X);
D = X - ones(N,1)*m;   % convert X to deviations from mean
SS = D'*D;
                       % partition the matrix SS
```

```
S = SS(1:p-1, 1:p-1);
P = SS(1:p-1,p);
t = SS(p,p);
                    % solve for regression on the first variable
b1 = P(1)/S(1,1);
b0 = m(p) - b1*m(1);
r  = P(1)/sqrt(t*S(1,1));
SR1 = r*r*t;
fprintf('\nAnalysis of Variance Table.\n\n');
fprintf('Source              d.f.    Sums of Squares\n');
fprintf('----------------------------------------------\n');
fprintf('Regression on x1   1      %8.2f\n',SR1);
SRL = SR1; %last sum of squares due to regression
for i = 2:p-1
   SI = S(1:i,1:i);    %redefine S,B,SR
   PI = P(1:i);
   BI = SI\PI;
   SRI = BI'*PI;       %sum of squares due to this regression
   SRA = SRI - SRL;    %sum of squares added by this regression
   fprintf('Added by x%g        1      %8.2f\n',i, SRA);
   SRL = SRI;
end
fprintf('Residuals          %2.0f      %8.2f\n',N-p,t-SRI);
fprintf('----------------------------------------------\n');
fprintf('Total              %2.0f      %8.2f\n\n',N-1,t);
b = BI;
SR = SRI;
```

Before running `mreg`, the original file matrix `plates.dat` must be reorganized so that the columns are in the proper order of decreasing correlation. This is done by using

```
X = [plates(:,4) plates(:, 2) plates(:,3) plates(:,1)...
     plates(:,5)]
```

Alternatively, a neater way of carrying this out is to define a vector `k = [4 2 3 1 5]` and rearrange the columns of `plates` using `X = plates(:,k)`. The results printed out by `mreg` are shown in Figure 5.5. The sum of squares added by regressing on continental area (the new x_2) in addition to effective trench length (the new x_1) is 24.67. The total sum of squares remain the same no matter how many independent variables are used, so the error sum of squares becomes $SSE = 97.98 - 51.57 - 24.67 = 21.74$. It has 9 degrees of freedom, so the mean square is 2.42. The F value obtained by dividing the mean square for the second variable alone (24.67) by 2.42 is $F = 10.2$. This value is still larger than the tabulated F value for 1 and 9 degrees of freedom (5.12)—so we conclude that adding this variable improves the regression. The sums of squares added by the remaining two variables are small, so it is clear that adding them does not further improve the regression.

```
>> [b,SR,t] = mreg(X)
Analysis of Variance Table.

Source               d.f.     Sums of Squares
------------------------------------------------
Regression on x1      1             51.47
Added by x2           1             24.67
Added by x3           1              2.39
Added by x4           1              5.70
Residuals             7             13.74
------------------------------------------------
Total                11             97.98
```

Figure 5.5   Analysis of variance for plates example

mreg reports the coefficients of the regression line; they are needed for computing predicted values of y, but it is very difficult to interpret these coefficients. In any event, they change value as each new variable is added. The square of the multiple correlation coefficient is more meaningful: it indicates (as r^2 does) the proportion of the total sums of squares "explained" by the independent variables.

We must note that correlation does not really imply causation: two variables may be correlated because they are both functions of some third, more fundamental but perhaps unmeasured variable. Is it reasonable to suppose that plates move because they are dragged along by subduction, and the speed is reduced for plates that have continents, because the thicker plate below continents causes extra drag on the base of the plate? Most geophysicists agree with Forsyth and Uyeda (1975) that this is plausible. The reasons have little to do with statistical testing—none of the test results reported here were given in the original paper! As Moss et al. (1995, p.1013) remark

> ... it is not sufficient to take statistical results at face value without some input of [geological] common sense.

Finally, among many possible questions that one might ask about this application, is this one: to what extent does it make sense to regard measurements on modern plates as random samples of some larger population? What is the sampled population? The answer makes use of a common assumption made in the earth sciences: what is happening at one instant of time ("the present") is representative of what has happened many times earlier (in "the past"). This is a crude version of what statisticians call "the ergodic hypothesis:" the property of many time-dependent processes that the "steady state" of the system does not depend on the initial state, or the time that the system has been operating. If this is so, the state, constantly changing over time, may be regarded at any particular time as a random sample taken from all possible states of the system. So if we assume that the action of plate tectonics has been more-or-less the same, at least since the Mesozoic (the last 200 million years or so: a plausible, limited form of uniformitarianism), then present plate configurations may be treated as simply a random sample of the many plate configurations that have existed since that time. Whether or not this is true is certainly not a question that the statistician can answer.

In MATLAB multiple linear regression can also be calculated using the backslash operator, just as it can be for simple linear regression. We need to modify the function given in the previous section only slightly.

```
function b = mlreg(X,y)
% b = mlreg(X,y)
% linear regression using the backslash operator
% X is a matrix, with the N data (rows) for
% m variables (cols). b is a vector: b(1) = intercept,
% the other b(2:m+1) are the coefficients for each
% variable
N = length(y);
X = [ones(N,1), X];
b = X \ y;
```

5.5   Polynomial Regression

It is fairly straightforward to extend the techniques of multiple linear regression to fitting a polynomial curve to data. For a single independent variable x, the polynomial regression model is

$$y = b_0 + b_1 x + b_2 x^2 + \ldots + b_m x^m + \epsilon \qquad (5.19)$$

To use this model, therefore, we simply compute the squares, cubes, etc. of the original x values, and treat these as though they were independent variables. Then we can use the techniques of multiple linear regression, in the same way that we did in the previous section. It is even simpler than ordinary multiple regression because there is now a logical way to add independent variables: we start with only the linear term, then add the quadratic term, then the cubic term, and so on, until adding higher terms no longer significantly improves the regression. (In rare cases, it may happen that the quadratic is no improvement on the linear, but the cubic is—as usual, it helps to plot the data and think about whether this might possibly be true.)

We give the following example of an application of polynomial regression. In a classic study of the stable isotopes of oxygen and hydrogen in the deep formation waters of sedimentary basins, Clayton et al. (1966) present a set of data determined on 26 samples from the Michigan basin: analyses were made for δD (deuterium, the heavy isotope of hydrogen, measured in percent difference from the ratio deuterium/hydrogen in standard seawater), δO^{18} (the heavy isotope of oxygen, measured in parts per mil difference from the ratio O^{18}/O^{16} in standard seawater), temperature T (measured in degrees Celsius), and total dissolved solids. These data are provided in the ASCII file mibas.dat.

Clayton et al. (1966) interpreted the formation waters as meteoric water, whose composition had been modified with reactions with the aquifers (composed largely of carbonate minerals). Because temperature affects isotopic fractionation, we expect that there should be a relationship between δO^{18} and T. It can be shown theoretically that this relationship should be nonlinear. Figure 5.6 shows the result of fitting a linear and quadratic curve to

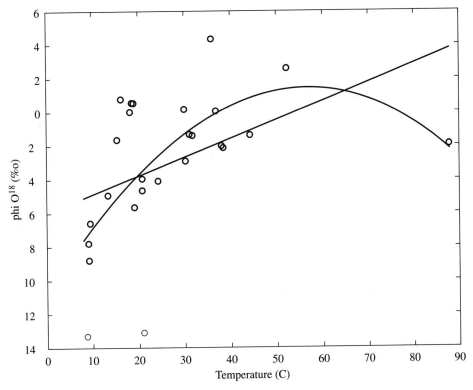

Figure 5.6   Linear and quadratic regression of δO^{18} against temperature, for groundwater in the Michigan Basin (data from Clayton et al., 1966)

the data, using the MATLAB function `polyfit`. Because it is not entirely straightforward how this is done, we list the steps in the following script.

```
load mibas.dat
x = mibas(:,3); % T
y = mibas(:,2); % delta O18
xmin = floor(min(x)); % minimum T, rounded to integer
xmax = ceil(max(x)); % maximum T, rounded to integer
xx = (xmin:xmax);          % T integers from xmin to xmax
p1 = polyfit(x,y,1); % linear fit
p2 = polyfit(x,y,2);       % quadratic fit
yl = polyval(p1, [xmin; xmax]); % evaluate linear fit
yq = polyval(p2, xx); % evaluate quadratic fit
plot(x,y,'o'); % plot data
hold on;
plot([xmin;xmax], yl); % plot linear fit
plot(xx, yq); % plot quadratic fit
```

Using `mreg` shows not only that the linear fit is significant, but also that adding a quadratic term produces a significant improvement in the fit, as can be seen from the following analysis of variance table. The observed F value is 6.98, compared to a tabulated value of 4.28 for $\alpha = 0.05$ and 1 and 23 degrees of freedom. Comparison of the quadratic fit with the theoretical line shown in the original paper demonstrates a remarkably good correspondence between the theoretical line and the statistical fit. Examination of Figure 5.6 might suggest that the quadratic fit is unduly influenced by one point, but re-running the program with this point removed shows that this is not really the case.

source	d.f.	sums of squares	mean squares	F
Linear regression	1	88.58	88.58	7.17
Quadratic	1	86.27	86.27	6.98
Error	23	284.37	12.36	
Total	25	459.22		

Polynomial regression may be extended to deal with polynomials of two or more independent variables. The case where the two independent variables are geographic coordinates is particularly important in the earth sciences, and is known as *trend analysis*. We defer treatment of this important application to Chapter 8, where we considered it along with other techniques of representing areally distributed data, such as contouring.

5.6   Reduced Major Axis

One of the problems about applying regression in the earth sciences is that the model assumptions are rarely valid for observational data. Remember that the independent variable x is supposed to be measured without error. Generally earth scientists are interested in fitting lines to plots of y against x where *both* variables are subject to "error," whether that error is experimental or due to natural variation. Often there is no clear distinction to be made between independent and dependent variables: which is plotted on the x-axis, and which on the y-axis is arbitrary or a matter of convention. Examples include chemical or isotopic analyses, or two morphological measurements, such as the height and width of a fossil shell, or the area and relief of a drainage basin. One of the advantages of the correlation coefficient as a measure of association is that it does not depend on any distinction between dependent and independent variables. For many applications, however, we require a fitted line.

This problem has no easy theoretical solution: for an optimum fit more information is required than is generally available. But there is a simple solution, which is often adopted. Instead of minimizing the departures of y from the estimated values (or alternatively, the departures of x from the estimated values) we can minimize the normal distance of the points (x, y pairs) from the fitted line. The line is then called the *reduced major axis*. Its slope is very readily calculated as $b = SS_y/SS_x$. For more details, see Davis (1986, p.200–204), and Troutman and Williams (*in* Size, 1987).

Kent et al. (1990) show how the multiple regression of y on x_1 and x_2 may be estimated, without assuming that all the error is in the dependent variable, if the variance-covariance matrix is known. The solution is obtained by successive approximations, using ordinary multiple regression to give the first estimates of the parameters. This case is important for fitting lines to isotopic data for age dating: for example, the dependent variable might be the ^{40}Ar abundance, and the two independent variables the ^{36}Ar and ^{39}Ar abundances. The technique used is very similar to that described in the next section (see Albarède, 1995, p.294–307, for further discussion and a numerical example from lead isotope dating).

5.7 Nonlinear Regression

In a previous section, least squares techniques were used to fit a polynomial function to data. One might surmise that this is a form of nonlinear regression. Indeed it is, but although the fitted function is nonlinear, the normal equations are linear in the fitted parameters (the b_i). This is not what is called nonlinear regression: in nonlinear regression, the problem is that the normal equations are nonlinear in the parameters (in this case, generally designated β_i)—and the nonlinearity cannot be removed by transforming the original x (or y) scale. As an example, consider the following two exponential functions:

$$y = \exp(\beta x) \tag{5.20}$$
$$y = \beta_0 \exp(\beta x) \tag{5.21}$$

Taking logarithms transforms the equations to

$$\log y = \beta x \tag{5.22}$$
$$\log y = \log \beta_0 + \beta x \tag{5.23}$$

The first equation is linear in the parameter β, but the second equation is not linear in the parameters β_0 and β. So if we have to use least squares to fit the second equation to a set of measured x and y values, then we have the problem that the normal equations are not linear equations. We can easily derive them, by taking partial derivatives;

$$\frac{\partial y}{\partial \beta_0} = \exp(\beta x) \tag{5.24}$$

$$\frac{\partial y}{\partial \beta} = \beta_0 b \exp(\beta x) \tag{5.25}$$

The second of these two equations cannot be made linear in β_0 and β by any transformation.

This means that nonlinear regression is more complicated than linear regression: we must first derive the normal equations, then solve them by a method of successive approximations. A common method uses an initial estimate (guess) at a plausible set of parameters (β_i^0), (the superscript indicates this is just the first guess), then derives new values of the β_i using a linearized (Taylor series) version of the normal equations, in the region of the estimated parameters. But one must be aware that if the original guess is not good, then successive approximations may diverge, or give unrealistic results.

Suppose, for simplicity, that the nonlinear function to be fitted is a function of only two parameters β_i, i.e., $y = f(x, \beta_1, \beta_2)$. Then using a Taylor series expansion, with only linear terms retained, we can write it as

$$y = f(x, \beta_1^0, \beta_2^0) + \frac{\partial f}{\partial \beta_1}(\beta_1 - \beta_1^0) + \frac{\partial f}{\partial \beta_2}(\beta_2 - \beta_2^0) \tag{5.26}$$

which can be written more generally in matrix form as

$$\mathbf{y} - \mathbf{f}^0 = \mathbf{A}_0 \mathbf{b}^0 \tag{5.27}$$

where $\mathbf{A}$ is the $N \times 2$ matrix of partial derivatives evaluated for each of the N x data and for each of the two parameters β_i, and $\mathbf{b}$ is the vector of differences between the fitted values of β_i and the initial guesses β_i^0. This matrix equation is easily solved using MATLAB. We then use the new estimates of the β_i to repeat the process until the difference between the new and old estimates is less than some value Δb.

The following example shows a MATLAB function which carried out these calculations for an example taken from Draper and Smith (1980, Chapter 10). In this case the equation to be fitted to a set of measurements of "available chlorine fraction" y_j as a function of time t_j (in weeks) is known to have the form

$$y = f(t) = \beta_1 + (0.49 - \beta_1) \exp(-\beta_2(t - 8)) \tag{5.28}$$

so the partial derivatives are

$$\frac{\partial f}{\partial \beta_1} = 1 - \exp(-\beta_2(x - 8)) \tag{5.29}$$

$$\frac{\partial f}{\partial \beta_2} = -(0.49 - \beta_1)(x - 8) \exp(-\beta_2(x - 8)) \tag{5.30}$$

The solution is obtained using the following function:

```
function beta = nlreg(x,y, beta0)
% function beta = nlreg(x,y, beta0)
% performs nonlinear least squares fit of
% f(x) to y by Gauss-Newton iteration.
% beta0 is a column vector of first guesses
% for beta. The appropriate nonlinear function func
% and pdfunc for calculating the partial derivatives
% of f wrt the betas must be written into the function.
% Written by Gerry Middleton, December 1996
% from examples in Albarede (1995, p.273-276) and Draper
% and Smith (1980, p. 458-481).
beta = beta0;
```

```
% first define the functions
yy = y;
z = beta(2)*(x-8);
func = 'beta(1) + (0.49 - beta(1))*exp(-z)';
pdf1 = '1 - exp(-z)';
pdf2 = '-(0.49 - beta(1))*(x - 8).*exp(-z)';

% do first approx
f = eval(func);
A(:,1) = eval(pdf1);
A(:,2) = eval(pdf2);
betad = A \ (yy-f);
beta = beta + betad
n = 0;

% now loop
while (max(betad) > 0.001)
   z = beta(2)*(x-8);
   f = eval(func);
   A(:,1) = eval(pdf1);
   A(:,2) = eval(pdf2);
   betad = A \ (yy-f);
   beta = beta + betad
   n = n+1;    % break out of loop after 50 iterations
   if n > 20
      break
   end
end
xx = x;            % save original x scale
xmin = min(x);   % make new one
xmax = max(x);
xd = (xmax-xmin)/50;
x = [xmin:xd:xmax];
z = beta(2)*(x-8);  % evaluate function for each x
f = eval(func);
plot(xx,y,'+',x,f)  % plot original pts and function
```

The result of running this function using the data in dstcl.dat and the initial guess [0.30;0.02] for beta0 is shown in Figure 5.7. Note that not all initial guesses give equally good fits. For an example taken from oceanography, see Albarède (1995, p.273–276).

A better method of approximation, known as the *Marquardt-Levenberg* method is described by Draper and Smith (1980), Bevington and Robinson (1992), and Press et al. (1989, Chapter 14). It is implemented in a function leastsq in the MATLAB Optimization toolbox, or by leasqr available from The MathWorks, Inc. ftp site.

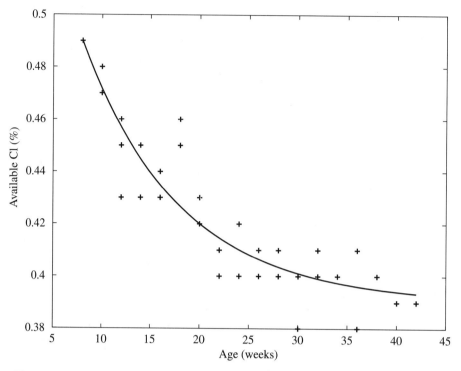

Figure 5.7   Nonlinear regression fit to data of an industrial experiment, taken from
Draper and Smith (1980).

Another method, generally called the Simplex method, is described by Caceci and
Cacheris (1984; see also Press et al., 1989, Chapter 10). It is implemented for MATLAB
by the function simplex (which uses rsum and simplot—all these M-files are included
in the accompanying diskette). It has the advantage of being very robust, though it may
converge more slowly on the true solution than the Marquardt-Levenberg method. To apply
this method to the dstcl data, we must first define the function we are going to fit in an
M-file. For these data we use the function defined in betafn, then type in

```
msimp = simplex(dstcl, [0.3 0.02],'betafn')
```

The fit is good, and it is easy to verify that it is insensitive to the starting point.

5.8   Splines

Some data may be considered to be measured almost without error. For example, in size
analysis by sieving or settling tube techniques the measured specimen consists of several
million grains. Size frequency is measured as weight percent between chosen size "grades,"
(e.g., 0.5 to 1.0 mm, or 1 to 0 phi), and the reproducibilty of replicate analyses is generally

about one percent of the weight. We know that the observed cumulative curve is not likely to be a straight line or a polynomial, no matter how we transform the original variables. So in constructing a curve, we should not use regression analysis, but a fitting technique that ensures that the fitted curve passes exactly through every measured point.

Draftsmen do this by using either "French curves," or "splines," which are thin, flexible rulers that can be bent around pins inserted at the data points. The mathematical technique of fitting splines is closely related to the draftman's physical spline. It is not the only interpolation technique in common use (another is the use of "Bezier curves," originally developed to represent smooth curves used in automobile design) but it is the commonest and simplest technique—and it is also readily available in a MATLAB implementation.

Before considering spline functions, let us first consider what makes a curve $f(x)$ "smooth?" In general, we require

- continuity of $f(x)$—there must be no breaks in the curve. This is easily achieved by drawing straight line segments between the points.

- continuity of derivative of $f(x)$. The slope of fitted straight lines changes abruptly from one segment to the next, so the first derivative is not continuous. At the least, therefore, we require continuity of first derivatives, and for truly smooth curves perhaps also of second derivatives.

In fitting a curve to data points we may be seeking a line that passes through several tens (or even hundreds) of points. It is not practical to use a single polynomial for this purpose: a straight line segment is defined by two points, a quadratic polynomial by three points, and a cubic polynomial by four points. So to fit many points we would need a very high order polynomial. Instead, we use a "piecewise" fitting technique: just as we can fit straight line segments between each pair of points, so we can fit a higher order polynomial, commonly a cubic fitted to two points, with the added condition that the segments must fit smoothly together, so the first and second derivatives must be equal at all the joints between fitted segments. A variety of choices are possible, which makes the mathematics of splines a fairly complicated subject, treated at length in texts on numerical analysis (e.g., see Lindfield and Penny, 1995). In what follows we describe just one of these, the option implemented by MATLAB.

First note that a cubic polynomial has at most three derivatives

$$
\begin{aligned}
y &= b_0 + b_1 x + b_2 x^2 + b_3 x^3 \\
y' &= b_1 + 2b_2 x + 3b_3 x^2 \\
y'' &= 2b_2 + 6b_3 x \\
y''' &= 6b_3
\end{aligned}
$$

The first derivative is the *slope* of the tangent to the curve at that point, and the second derivative is commonly called the *curvature* at that point. The third derivative is a constant for a cubic spline of known cubic coefficient b_3.

Now consider the simple problem of fitting four cubic splines to five points. The problems are different for the beginning and end points, $(x_1 y_1)$ and $(x_5 y_5)$, and the three interior points. At each of the three interior points (called "knots" in the technical literature on splines), we specify the position y, the first derivative y', the second derivative y'', and the third derivative y'''—but these conditions are shared between two curve segments, so instead of having eight conditions for the two curve segments defined by the three interior points, we only have four, two for each pair of points. This is what we need to define each curve segment uniquely. At the beginning and end points, however, we know what the position is, but we generally do not know any of the derivatives. So a choice of conditions is necessary: MATLAB chooses to use the same polynomial in the first segment x_1x_2 as in the second segment x_2x_3 (and similarly for the last and next to last segments). In fitting the first two segments, we have two positions specified, and the condition that, at the third point the position, and three derivatives are also specified, but shared with the next segment: so once again we have four constraints, which is what is necessary to define a cubic curve uniquely. This choice (an arbitrary one) is called the "not a knot" condition for the beginning and end points.

Another commonly used condition, discussed by Davis (1986) is the *relaxed* or *natural* end condition, which assumes (equally arbitrarily) that the second derivative (and therefore also the third) is zero at the first and last points—in other words the slope is not changing at those points. The name derives from comparison with the draftsman's spline: at the first and last pin, the curvature of the flexed ruler is zero, because no force is being applied to bend it around those pins.

Davis (1986) and Lancaster and Šalkauskas (1986) give more details about computation. This involves generating large sets of linear equations and solving them. The main matrix is "sparse," meaning that most of the entries are zero, which makes it possible to use efficient solution techniques. We need not be concerned with the details here. Instead we give an example application using MATLAB (for other MATLAB applications, see Hanselman and Littlefield, 1998, Chapter 20).

The original data (in the file bedf.dat) consists of a part of a profile, measured over sand ripples using an accurate depth sounder in a flume. The points are spaced 7 mm apart, so linear interpolation (as performed by MATLAB's plot function) almost gives the illusion of a smooth curve, though there are irregularities probably produced by small instrument errors (and magnified by the $\times 4$ vertical exaggeration used to plot the data: Figure 5.8). As an exercise, we show the result that would be obtained by using spline to interpolate using only every fifth data point. The appropriate script follows:

```
% script to fit spline to bedform data
load bedf.dat;              % 121 elevations in cms
x = [0:0.7:84];             % horizontal scale in cm
xx = x(1:5:121);      % reduced set -- every fifth point
yy = bedf(1:5:121);
yi = spline(xx, yy, x);     % spline fit, evaluated at x
subplot(2,1,1);     % upper part of figure
plot(x, bedf, xx, yy,':');
subplot(2,1,2);     % lower part of figure
plot(x, yi, xx, yy,':');
```

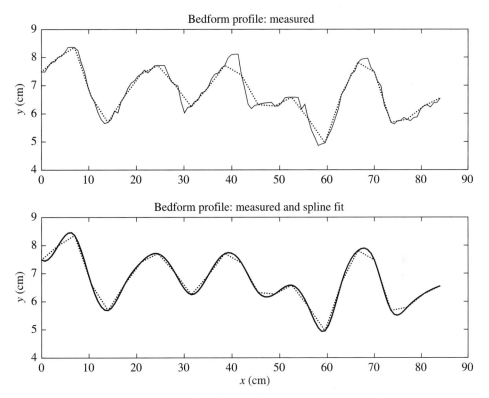

Figure 5.8   Upper box shows profile of several sand ripples, measured by accurate sonic depth sounding at intervals of 7 mm in a laboratory flume (full line). Profile produced by fitting line segments to every fifth point is shown dotted in both boxes. Lower box shows profile fitted by spline interpolation.

Figure 5.8 illustrates not only the use of MATLAB's `spline` function, but also `subplot`. This function can be used to produce multiple plots within a single figure. See the MATLAB manual for details.

One of the disadvantages of spline fits is that the curve produced generally looks "unnaturally" smooth. Curves describing real-world objects or functions often have bends with sharper curvature than simple spline fitting will allow. A modified spline technique, called "splines under tension" has been developed to simulate sharp bends (for an example and references see Middleton, 1990).

5.9   Recommended Reading

Albarède, F., 1995, Introduction to Geochemical Modeling. Cambridge University Press, 543 p. (QE515.A53 Chapter 4 discusses Probability and Statistics, and Chapter 5 discusses inverse methods, including nonlinear least squares. This book treats a wide variety of different types of models, using matrix algebra, and gives many example calculations carried out using MATLAB – but the author lists no MATLAB scripts.)

Bevington, P.R. and D.K. Robinson, 1992, Data Reduction and Error Analysis for the Physical Sciences. New York, McGraw-Hill, Second edition, 328 p. (QA278.B48 Good treatment of regression and curve fitting: paperback with a diskette of Pascal programs.)

Caceci, Marco S. and William P. Cacheris, 1984, Fitting curves to data: The Simplex algorithm is the answer. Byte, May 1984, p.340–362. (Gives a good discussion of the method, and a Pascal program from which `simplex` was derived.)

Cox, Allan and R.B. Hart, 1986, Plate Tectonics: How It Works. Palo Alto, Blackwell Sci., 392 p. (QE511.4.C683: an elementary yet thorough introduction.)

Draper, N.R. and H. Smith, 1981, Applied Regression Analysis. New York, John Wiley and Sons, Second edition, 709 p. (A very thorough treatment.)

Lancaster, P. and K. Šalkauskas, 1986, Curve and Surface Fitting: An Introduction. New York, Academic Press, 280 p. (QA297.6 A concise summary of many techniques valuable to earth scientists: cubic splines are discussed in Chapter 4.)

Lindfield, G. and J. Penny, 1995, Numerical Methods Using MATLAB. New York, Ellis Horwood, 328 p. (QA297.P45 Provides a more advanced mathematical discussion of many of the topics covered in this chapter.)

Shaw, D.M. and J.D. Bankier, 1954, Statistical methods applied to geochemistry. Geochimica et Cosmochimica Acta, v.6, p.111–123. (An excellent discussion of t- and F-tests applied to regression and differences in means and variances between samples.)

Chapter 6

Classification

6.1 Introduction

Classification is a fundamental problem for science: what are the fundamental particles and their properties? what are the chemical elements? what are species, and how can they (and higher taxa) be defined for living and fossil organisms? how can natural environments or ecosystems be identified and distinguished from one another? These are all examples of problems that have or still do concern scientists. Solutions cannot be found only by using numerical data—the full, complex, interplay of scientific concepts (theories, hypotheses), observations and measurements must be used. Measurements, at all levels (nominal to ratio), can nevertheless be used to make classifications more precise, and easier to use for non-experts. Many students beginning the study of mineralogy, for example, may be frustrated to see the ease with which an expert mineralogist can identify mineral specimens. The expert may not even be able to explain just how he knows, for example, that a particular specimen is the mineral, scapolite. The classification of minerals, however, is now in such an advanced state that even a student can identify most minerals by making a few semi-quantitative measurements.

Classification, in the broad sense of the word, refers to two quite different procedures:

- **Assignment**, that is, placing unassigned objects in one of a known set of classes. Assignment is carried out by seeking the class that contains objects more similar to the unassigned object, than the objects in any other class.

  This is a simpler problem than devising a new set of classes. In some fields of study, the classes are now well-known, e.g., the elements, and most chemical compounds, the minerals, and most of the living species in higher animal groups. It is generally unnecessary to devise a new classification, or even to modify the existing one. Assignment to known classes is the problem faced by the mineralogist who is asked to identify an unnamed mineral specimen, or by a paleontologist who has to classify an unnamed fossil in a well-studied group.

- **Classification** in the strict sense, that is, identifying natural classes by examining a large number of objects. Natural classes contain objects that are more similar to each other than the objects in all the other classes.

  In many fields of study the classes are still unknown or debatable, e.g., many fossil species and higher taxa, ecological communities, sedimentary environments, and geomorphic forms.

Scientists in disciplines with well-established classifications often discount the significance of classification: they tend to regard all classifications as arbitrary, and debates about classification as pointless. But it is not true that all classifications are arbitrary: it seems obvious to us now that there is an important distinction between elements and compounds, and that elements can be classified using the periodic table—but it was not at all obvious much earlier in the history of chemistry. Debates about classification are generally not simply about definitions ("semantic" debates) but about the deeper meanings that lie behind definitions. In particular, scientists generally prefer classifications that are theoretically based ("genetic"), rather than simply "descriptive."

In the area of taxonomy (the classification of organisms) the distinction between these two types of classifications has been made explicit. Classifications based on measurable properties of organisms (where "measurable" is defined broadly to include all the scales of measurement) are described as *phenetic*. Classifications based on the inferred evolution of organisms are described as *phylogenetic*. One form of phylogenetic classification is that based strictly on genealogy, and this is described as *cladistic*. The distinction between phylogeny and cladistics is, however, not simple (for a book-length treatment, see Panchen, 1992).

Clearly, all scientists would like the classifications they use to have two properties: (i) they should permit clear and unambiguous assignment of unknowns to known classes (where possible, or to a class of "none of the known classes", where the known classes are inappropriate); and (ii) they should correspond to significant, well-established theoretical categories. In other words, they should be *both* descriptive *and* genetic. In a few cases, e.g., the periodic table of the elements, this is possible. In most cases, however, a genetic classification is also subjective, since it depends on theoretical judgments not shared by all experts in that field. In these cases, objectively descriptive classifications are generally established and used by common consent, even though it is known that they are somewhat arbitrary because their theoretical basis is inadequate. In the earth sciences, this is true of classifications of soils, sediments and rocks (petrographical names), of stratigraphic units, of physical, chemical, and biological environments, of geomorphological forms, and of many other classifications.

In this chapter, we are concerned only with descriptive, numerical classification: but most of the techniques were devised to make such classifications less arbitrary and more natural, while still being strictly objective in the sense that the same procedure carried out by another scientist leads to the same classification. Absolute objectivity remains unattainable because it is never clear which is the "best" procedure to be used.

6.2 Measures of Similarity

In Chapters 4 and 5 we defined the correlation coefficient as a measure of "association" of two standardized variables. The correlation coefficient may therefore be used as a scale-independent measure of similarity of two variables. Its use is easily extended to measuring the similarity of two objects, on each of which several variables have been measured. In this case we simply transpose the rows and columns of the data matrix, so that the columns now represent samples, rather than variables, and the rows represent variables, rather than samples. The correlation matrix may then be calculated in the usual way, but the coefficients relate samples (objects) rather than variables. The following is a MATLAB implementation:

```
Xt = X';  % X is a data matrix with variables as columns
Rq = corrcoef(Xt);
```

These two ways of calculating similarity are generally called *R-mode* (between variables) and *Q-mode* (between samples).

In calculating correlation coefficients we have made two subjective judgments: (i) Sampling is generally assumed to be random, but the choice of variables to be measured is never random. This does not matter in R-mode analysis, but choice of variables will greatly affect Q-mode analysis. (ii) The correlation coefficient was chosen as the measure of similarity.

In some cases, the correlation coefficient is clearly not the appropriate measure. This is the case if the variables are measured on the nominal (e.g, "absent," "present") or ordinal (e.g., "absent," "rare," "common," "abundant") scale. In such cases, a variety of other measures have been proposed, mainly for use in Q-mode analysis.

For a binary nominal scale, "absence" (or "not having some property") may be coded 0, and "presence" (or "having some property") 1. Let n be the number of variables studied, $n11$ the number of variables both coded 1 in two samples, $n01$ the number of variables coded 0 in the first sample, but 1 in the second, and so on. The number of "matches" is defined as $n11 + n00$. Then the *simple matching coefficient* between two samples is defined as

$$s_m = \frac{n11 + n00}{n} \tag{6.1}$$

and *Jaccard's coefficient* is defined as

$$s_j = \frac{n11}{n11 + n01 + n10} \tag{6.2}$$

It differs from the simple coefficient by excluding "negative" matches (cases where both samples *lack* some property). *Gower's similarity coefficient* is a measure that may be used when the data consist of measurements made on a mixture of ordinal and interval or ratio scales (see Dunn and Everitt, 1982).

Even when the measurements are all made on the interval or ratio scale, it is clear that the correlation coefficient is not always the most appropriate measure of similarity between samples. A commonly used alternative is some measure of the distance between the two samples in n-dimensional variable space. The *standardized Euclidean distance* between two samples (1 and 2) is given by

$$d = \sqrt{\left(\sum_{k}^{n}(x_{1k} - x_{2k})^2\right)/n} \tag{6.3}$$

The following MATLAB function calculates a matrix of these distances for a data matrix X (with N rows and n columns):

```
function D2 = edist(X)
% D2 = edist(X)
% calculates the standardized Euclidean distance for the
% N samples in the rows of the data matrix X
% written by Gerry Middleton, October 1996
[r c] = size(X);
D2 = zeros(r);  % reserve space for D2
for i = 1:c
    X1 = X(:,i)*ones(1,r);  % ith col values in r cols
    DX = X1-X1';    % differences
    D2 = D2 + DX .^2;    % sum squared differences
end
D2 = D2/c;    % standardize
```

(For a version that does not use a for loop, see Appendix A.) The distance matrix is symmetrical, just as the correlation coefficient matrix is. The larger the distance, the smaller the similarity. The principal diagonal (the distance of a sample from itself) is zero, rather than one as in the correlation matrix. One of the main disadvantages of the Euclidean distance is that, unlike the correlation matrix, it depends strongly on the scaling of the variables used.

Another similarity measure, with stronger affinities to the correlation coefficient is the *Cosine Theta* measure of proportionate similarity. For two samples 1 and 2 this is defined as

$$\text{cosine } \theta = \frac{\sum_{k=1}^{n} x_{1k}x_{2k}}{\sqrt{\sum_{k=1}^{n} x_{1k}^2 \sum_{k=1}^{n} x_{2k}^2}} \tag{6.4}$$

If the variables are standardized as z-scores, Cosine Theta is identical to the correlation coefficient. If the variables are not standardized, Cosine Theta is sensitive to the relative proportions, not the absolute magnitude of the variables. This makes it particularly appropriate to compositional variables that measure the proportion of a given constituent in a sample (e.g., the percentage of coral fragments in a Bahama sediment). Measurements made to define the morphology of organisms generally consist of a large number of lengths (height, width, length of various anatomical parts, etc.). If the proportions between the lengths are the same in two specimens, this means that the specimens have the same shape, though the size may be different. Cosine Theta is one in this case, an appropriate value in taxonomy where size is generally considered much less important than shape. The following MATLAB function calculates a Cosine Theta matrix from a data matrix X:

Raw Data				Percent Data			
localities				localities			
1	2	3	4	1	2	3	4
50	25	60	50	50	50	43	25
30	15	40	50	30	30	29	25
20	10	30	50	20	20	21	25
00	00	10	50	00	00	7	25

Correlations				Cosine Theta			
localities				localities			
1	2	3	4	1	2	3	4
1.000	1.000	1.000	0.000	1.000	1.000	0.989	0.811
	1.000	1.000	0.000		1.000	0.989	0.811
		1.000	0.000			1.000	0.889
			0.000				1.000

Table 6.1   Upper two tables show hypothetical data matrices, from Imbrie (1963). Rows are variables, columns are sample localities. Lower two tables show similarity matrices calculated from the data above.

```
function C = costheta(X)
% C = costheta(X)
% calculates a Cosine Theta similarity matrix from
% a data matrix X
% written by Gerry Middleton, October 1996
[r c] = size(X);
C = zeros(r);
RS = sqrt(sum(X' .^2));
D = RS'*ones(1,c);
W = X ./ D;
C = W*W';
```

RS is the denominator of Equation 6.4. Since each row of X has to be divided by the corresponding item in this vector, we have to copy it into the rows of D before using MATLAB's ./ operator to obtain W. The algorithm is described by Reyment and Jöreskog (1993: see also Davis, 1986, or Howard, 1991).

Imbrie (1963) provided the example shown in Table 6.1 to illustrate the advantage of Cosine Theta in describing similarity of samples (as opposed to variables, for which the correlation coefficient is generally more appropriate).

The correlation coefficients calculated from the raw data are all 1.00, except for the correlations for locality 4, which are all zero. This does not seem appropriate because the proportions of the different variables are identical for localities 1 and 2, but different for locality 3. The Cosine Theta values are 1.00 for localities 1 and 2, 0.989 for locality 3 with either 1 or 2, and 0.811 for locality 4 with either 1 or 2.

6.3   Assignment Using Discriminant Functions

Discriminant functions were introduced by the statistician R.A. Fisher to solve taxonomic problems (assignment of skulls to different anthopological classes, *in* Barnard, 1935; or assignment of botanical specimens to different species of iris, Fisher, 1936). The function that best discriminates between two classes is first set up using sets of samples that are known to belong to one or the other group. Suppose we have N_1 samples from the first class, and N_2 from the second. m variables have been measured on each of the samples. Then we can set up a "dummy" dependent variable y which is assigned the value $y_1 = N_2/(N_1 + N_2)$ for each sample of the first group and $y_2 = -N_1/(N_1 + N_2)$ for each sample of the second group. (For equal sample sizes $y_1 = 0.5$ and $y_2 = -0.5$.) Then we determine the multiple regression equation that best predicts y from the m measured variables x_i. This equation is the "discriminant function" that can then be used to assign an unknown sample to one or the other group: if the measurements give a positive y value it is assigned to the first group, and if a negative value it is assigned to the second group. The significance of the original discrimination between the two groups can be tested by using analysis of variance to test the significance of the regression on the "dummy" variable y. It should be noted that this is not an obvious application of multiple regression, because the dummy dependent variable has assigned, not measured, values; but it works nevertheless.

The geometry of discrimination can readily be illustrated for the case where only two variables have been measured. Figure 6.1 is a simple scatter plot of the two variables, produced by the following script.

```
function m = discpl2(x1,y1,x2,y2)
% m = discpl2(x1,y1,x2,y2)
% produces a two-dimensional plot
% of the two sets of x,y data contained
% in the column vectors x1,y1,x2,y2, using
% deviations from the means m(1) and m(2)
% of all the data
x = [x1;x2];    % combine data sets
m(1) = mean(x);    % and calculate deviations from mean
x1 = x1 - m(1);
x2 = x2 - m(1);
y = [y1;y2];
m(2) = mean(y);
y1 = y1 - m(2);
y2 = y2 - m(2);
c1 = length(x1);
c2 = length(x2);
d1 = c1/(c1+c2);  % calculate dummy variables
d2 = c2/(c1+c2);
z1 = d1*ones(c1,1);
z2 = -d2*ones(c2,1);
```

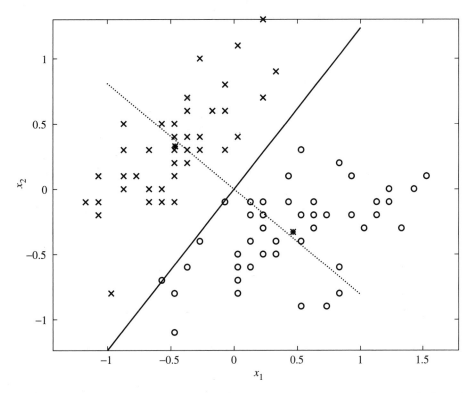

Figure 6.1   Scatter plot for two variables x_1 and x_2, measured on two groups, indicated by circles and crosses. Each variable shows considerable overlap between the groups, but the overlap can be reduced by separating the groups using the line shown. The data are measurements made on the flowers of two iris species, taken from Fisher (1936).

```
plot(x1,y1,'x',x2,y2,'o','LineWidth',1);
xlabel('x'),ylabel('y');
axis('equal');
hold on;
X = [x-m(1),y-m(2),[z1;z2]];  % calculate
[b,SR,t] = mreg(X)   % discriminant function
xx = [-1 1];
yy = (b(1)/b(2))*xx;
yy2 = (-b(2)/b(1))*xx;
x1m = mean(x1);
y1m = mean(y1);
x2m = mean(x2);
y2m = mean(y2);
plot(xx,yy,':')
plot(xx,yy2,x1m,y1m,'*',x2m,y2m,'*','LineWidth',1)
```

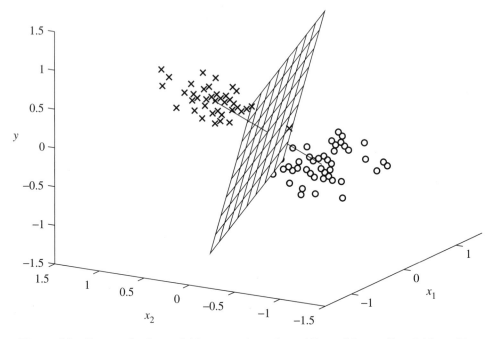

Figure 6.2   Scatter plot for variables x_1 and x_2 after adding a "dummy" variable y. The discriminant function is a plane $y = b_1x_1 + b_2x_2$. This plane cuts the x_1x_2 plane along the dotted line shown in Figure 6.1. The plane shown in the figure is normal to the plane of the discriminant function.

Two lines are plotted: the dotted one is actually the discriminant function, the full line is a line normal to that function, which passes through the grand mean, and best separates the two groups of data.

The data (in files `iriset.dat` and `irisvers.dat`) are the original botanical data analyzed by Fisher (1936), except that we have used only two of the four measured variables, and have normalized the measurements so that the x_i are deviations from the mean of both groups. Each variable shows considerable overlap between the two groups, but the overlap is essentially eliminated by using a combination of the variables. The dotted line has an equation of the form $x_2 = b_1x_1$, or equivalently $x_2/x_1 = b_1$: this mathematically expresses the fact that the best discriminator between the two groups is a ratio of the variables, rather than the value of either variable by itself. The value of the ratio that best discriminates the groups is b_1. Alternatively, we can use a line normal to the discriminant function (the full line in Figure 6.1) to separate the two groups of points. In this example, there are no samples misclassified: but that is not always the case. The aim is to reduce the number of samples misclassified below that expected from a random assignment (50% misclassified, in this example).

In Figure 6.2 the plot shows the same data plotted in three dimensions, using the "dummy" variable, assigned a value 0.5 in group 1 and -0.5 in group 2. The figure also shows a line joining the means of the two groups, and the plane that best separates the two groups, calculated from the discriminant function. The intersection of this plane with the $z = 0$ plane is the full line shown in Figure 6.1. This plot was produced by the function `discplot`, included on the diskette: it may be examined for further details.

The analysis of variance obtained using `mreg` is as follows:

source	d.f.	sums of squares	mean squares	F
Regression on x_1	1	13.26	13.26	313.2
Added by regression on x_2	1	7.63	7.63	180.2
Error	97	4.11	0.042	
Total	99	25.00		

It is clear that the F values are highly significant, indicating that the discrimination is significant, and that both variables contribute significantly to it. In this case, the conclusion is obvious from inspecting Figure 6.1—but in other applications, using more than two variables, it may not be so clear from inspecting scatter plots alone.

Many applications are more complicated than this, because there are more than two classes in the classification. Not all texts use the multiple regression approach outlined above, and there are many other complications that we have not considered (e.g., estimating the probability of misclassification, allowing for different *a priori* probabilities of different classes). For worked numerical examples and further discussion see Davis (1986), Dunn and Everitt (1982), Fisher (1936) and Le Maitre (1982). Anderson (1984) gives a thorough mathematical treatment. Examples in the earth sciences have been given by Potter et al. (1963), Shaw and Kudo (1965), Roser and Korsch (1988), Molinaroli et al. (1991) and Fedikow et al. (1991).

In actual applications, the main problem is not generally the technique itself, but the validity of the assumptions. As an example consider the many efforts that have been made to distinguish sands (and sandstones) deposited in different sedimentary environments by using size analysis. The variables can be considered to be either moment measures calculated from the size analyses, or the actual weight percents in each of the different size grades. The classes often considered are (i) eolian sands, (ii) beach sands, and (iii) fluvial sands. Studies of modern sands in particular coastal regions readily establish significant discriminant functions for these three groups of sands (e.g., Moiola et al., 1974; Stokes et al., 1989), but application to other areas and ancient sandstones fails for the following reasons:

- The classes established must include all those found in the target population. If the target population is "all sands" then three classes (eolian, beach, fluvial) do not exhaust all the possibilities—what about deltaic sands, deep sea sands (turbidites), shallow marine (intertidal or subtidal) sands, and possible differences between coastal and desert eolian sands?

- The samples used to determine the discriminant function must be random samples drawn from the target population. If the target population is "all modern sands" then the discriminant function must be based on a random sample of such sands (which is *not* easily collected: samples from North America and Europe cannot be considered typical of the whole world!). If the target population includes ancient as well as modern sands, then ancient sands must also be sampled. Many ancient sands (and even more so, sandstones) have had their size characteristics significantly modified by post-depositional alteration (diagenesis).

Problems of this type (population definition and random sampling) are generally much more important than assumptions made about the multivariate Normal distribution of the variables, or equality of the covariance matrices of the various classes.

6.4   Principal Component Analysis

We now begin a consideration of the more difficult of the two aspects of classification discussed in Section 6.1: how do we detect natural groups from measurements, when we do not already know whether such groups exist, or how many there are? As an example, we may make use of the iris data described by Fisher (1936). He tells us that the four measurements were made on "...the flowers of fifty plants each of the two species...found growing together in the same colony..." Such measurements might easily have been made by a scientist who was not aware that there *were* two species (though it would be unlikely that he would happen to have 50 specimens of each species in a randomly drawn sample). Could our hypothetical botanist have used the measurements themselves to indicate that there were two distinct species?

In this case, we expect that if there are two (or more) species, the samples will tend to fall into two (or more) morphologically distinct groups, with relatively few intermediate types. We might be able to detect such groups by plotting the data on scatter plots like Figure 6.1. Indeed, even if we had not been able to use different symbols for the two species, we might have suspected that two species existed, because there seem to be two concentrations of points, with a region with very few points separating them. The problem is to detect such "clusters" of samples, when we have measured more than two or three variables. Alternatively, we might look for clusters in sample space, rather than variable space. In our example, we would be looking for the way the variables cluster in a 100-dimensional sample space. These two ways of looking at the problem correspond to R- and Q-mode analysis respectively.

One way this problem might be approached is to try to reduce the number of dimensions involved. In the iris example, we might guess that the four length measurements are likely to be strongly correlated. Such measurements can often be regarded as expressing two (or more) fundamental aspects of morphology, e.g., overall size, and some aspect(s) of shape. A similar reasoning was used by psychologists who devised "intelligence tests." They were not so much interested in measuring particular knowledge or skills, as they were in measuring some supposed "general intelligence," which was imperfectly measured by any one test. Their efforts gave rise not only to the modern (and still controversial) study of human intelligence, but also to the branch of statistics called *factor analysis*. A key step in factor analysis is *principal component analysis*.

Principal components are linear combinations of variables (in R-mode analysis) which have maximum variance, subject to two constraints: the total variance must remain the same, and the components must be uncorrelated. The first component has the largest possible variance, the second is the component that is uncorrelated with the first, and has the next largest variance, and so on until all the original variance has been accounted for. Thus a set of principal components contains all the (linear) information in the original set of variables, but packaged in a different way. If there are n original variables, it may be possible to reduce them to $m < n$ principal components, without any loss of the linear relationships between the original variables. More probably, n components are still required, but a small subset $m < n$ will be found to account for a large proportion of the variance observed in the original variables.

Using the iris example, the first component c_1 is the linear combination of the original variables

$$c_1 = v_1 x_1 + v_2 x_2 + v_3 x_3 + v_4 x_4 \tag{6.5}$$

that has the largest variance, subject to the constraint that the sum of squares of the weights v_i is equal to one:

$$v_1^2 + v_2^2 + v_3^2 + v_4^2 = 1 \tag{6.6}$$

Note that, because of the constraint, the v_i may be interpreted as direction cosines, i.e., the linear transformation from the x_i to the c_1 takes place by a rotation from one frame of reference to another which maximizes the variance of c_1. The second component c_2 is given by a similar formula (with a different set of weights). It has the largest variance of any variable *that is not correlated with c_1*.

These concepts can be shown graphically for the case of two variables. Figure 6.3 shows a scatter plot of the first two variables for *Iris versicolor*. Both have about the same variance, and there is clearly a correlation. Performing a principal component analysis yields the two component axes C1, C2 shown on the figure. The first principal component now has a much larger variance than the second. The two components are uncorrelated, and can be obtained by a rotation of the original axes X1, X2 to the position of the component axes.

Readers already familiar with linear algebra will recognize that principal component weights are none other than the *eigenvectors* of the correlation or covariance matrix, and that the proportion of the total variance that each eigenvector "accounts for" is given by the magnitude of the *eigenvalue*. Determining the eigenvalues and eigenvectors of real, symmetrical matrices is a familiar numerical problem that arises in many branches of science, and notably (for earth scientists) in the determination of the principal stresses and strains from an observed stress or strain matrix (e.g., see Middleton and Wilcock, 1994, or Ferguson, 1994). Because it is a familiar numerical problem, it is also very easy to carry out using MATLAB.

For readers not familiar with the concept of eigenvectors and eigenvalues, the following is a "quick-and-dirty" introduction, using the iris data as an example. We first combine the data for *I. setosa* and *I. versicolor* in a single data matrix X (the first 50 rows are the *I. setosa* measurements, and the second 50 rows are the *I. versicolor* measurements). Then we calculate the correlation coefficient matrix (using the MATLAB command R =

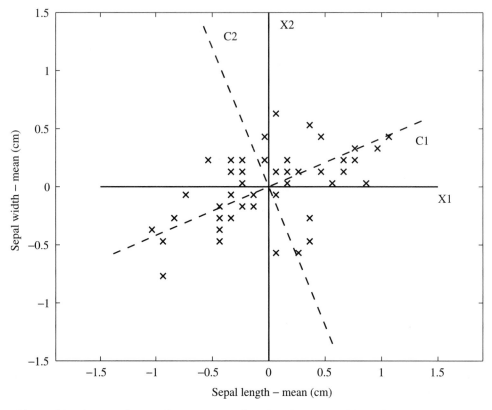

Figure 6.3   Scatter diagram for *Iris versicolor*, showing the position of the two principal component axes.

`corrcoef(X))` and obtain the result:

$$R = \begin{bmatrix} 1.0000 & -0.2059 & 0.8125 & 0.7896 \\ -0.2059 & 1.0000 & -0.6027 & -0.5709 \\ 0.8125 & -0.6027 & 1.0000 & 0.9793 \\ 0.7896 & -0.5709 & 0.9793 & 1.0000 \end{bmatrix}$$

This confirms what we suspected, namely that certain variables (3 and 4: petal length and width) are highly correlated, and both are correlated with variable 1 (sepal length). Variable 2 (sepal width), however, shows a negative correlation with the other three variables. It seems likely, therefore, that it should be possible to find linear combinations of the variables to obtain two new "components" that would account for most of these correlations: we expect that c_1 will be weighted on variables 1, 3 and 4; and c_2 will be weighted mainly on variable 2. We find these components by solving

$$(\mathbf{R} - \lambda_i \mathbf{I}) \mathbf{v_i} = 0 \qquad (6.7)$$

where the λ_i are the eigenvalues (variances of the principal components) and the $\mathbf{v_i}$ are the eigenvectors, that is, the weights that must be applied to the original data (converted to z scores) in order to obtain the principal components. We expect there will be 4 principal components, with 4 corresponding vectors of weights.

The solution to Equation 6.7 is obtained by finding the roots of the determinant

$$|\mathbf{R} - \lambda_i\mathbf{I}| = 0 \tag{6.8}$$

to obtain the eigenvalues. Once these are known, the corresponding eigenvectors can be obtained from Equation 6.7.

To obtain the eigenvalues, we have to find the roots to an order n polynomial, where n is the number of variables. Even for a fourth order polynomial, this is not an easy numerical task. Luckily, MATLAB makes it easy in practice: we simply enter the command

```
[V,D] = eig(R)
```

where R is the correlation matrix. V is the matrix of eigenvectors—each column gives the weighings for the corresponding principal component—and D is a diagonal matrix with the eigenvalues in the principal diagonal and zeros elsewhere. In the iris example, the result is

$$V = \begin{bmatrix} 0.5504 & 0.6739 & 0.1202 & 0.4781 \\ 0.8328 & -0.4015 & -0.0868 & -0.3710 \\ 0.0320 & -0.2907 & -0.7703 & 0.5667 \\ 0.0496 & -0.5479 & 0.6202 & 0.5592 \end{bmatrix}$$

and the corresponding eigenvalues are

$$\lambda_i = [\, 0.8068 \ 0.1302 \ 0.0167 \ 3.0463 \,]$$

Note that the sum of squares of any column in the V matrix is equal to one, and that the sum of the eigenvalues is 4.0. MATLAB does not order the eigenvalues by size, so the largest eigenvalue (traditionally designated λ_1) is here the fourth listed. As expected the fourth eigenvector is weighted on variables 1, 3, and 4. The second largest eigenvalue is the first listed, and the corresponding eigenvector is weighted mainly on variable 2, though also on variable 1. The sum of the eigenvalues is equal to the sum of the principal diagonal elements of the correlation matrix, which can be thought of as the sum of the variances of the standardized variates (z-scores). So the first principal component accounts for $3.05/4 = 76\%$ of the total standardized variance, and the second for a further 20%. The remaining two components can be neglected. This is very convenient, because it means we can use a two-dimensional scatter plot of the first two components to display the results. The steps in the calculation, described separately above, have been combined into the MATLAB function `princomp`, included on the diskette, which may be used to calculate principal components and plot the results for other sets of data.

To do this we first convert the data to z-scores (stored in the 100×4 matrix Z and then compute the two component vectors c1 and c2

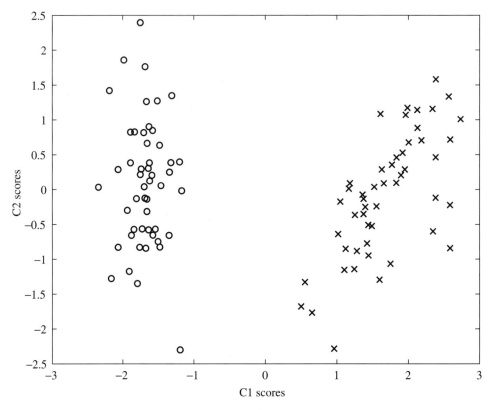

Figure 6.4   Scatter plot of iris data, using the first two principal components.

```
Xd = X - ones(100,1)*mean(X); % compute deviations from mean
Z = Xd ./ (ones(100,1)*std(X)); % divide by standard deviation
c1 = Z*V(:,4);
c2 = Z*V(:,1);
plot(c1(1:50),c2(1:50),'o', c1(51:100), c2(51:100),'x')
```

The plot is shown in Figure 6.4. Even without the use of two different symbols, it would be apparent that there are two different clusters shown. Note that the clusters and their separation are much clearer in this diagram than in Figure 6.1 which used only the first two original variables. This is because the first two principal components make use of most of the information contained in all four of the original variables.

The main disadvantages of principal components are: (i) They only make use of linear relationships between the variables—if the variables are related in a nonlinear way, then they should first be transformed to produce linear relationships, before using principal component analysis. (ii) The components are often difficult to interpret in physical terms. For example, the first component is weighted on three length measures, so it seems to be mainly a size component. But the second component is mainly weighted on sepal width, but also on sepal length (which shows a slight negative correlation with width, in the original data). It seems to be a mixed measure of both the size and shape of the sepal. The scatter plot shows that it is mainly the first component ("overall size") that distinguishes these two species.

In our example, we applied principal component analysis in R-mode to the correlation matrix. It can also be applied to the covariance matrix: the decision about which matrix to use depends largely on how significant one considers the units of the measurements to be. For example, if they are all length measurements (as in the iris example) one might use the covariance matrix. This would give larger weight to the longer measurements—which might not make taxonomic sense. Use of the correlation matrix is equivalent to using z-scores rather than unstandardized data: all the measured variables are assigned an equal weight. Principal component analysis can also be applied in Q-mode, and to similarity matrices using similarity measures that differ from the correlation coefficient (so long as the matrix remains real and symmetrical).

Factor analysis consists of a further simplification and transformation of principal components, generally by (i) neglecting some of the components, and (ii) rotating the others in space so that they are easier to interpret. The results are generally somewhat subjective, in that they depend on judgments about how many components can be neglected, and what constitutes the appropriate technique of "rotation for meaning" of the factor weighings. Since the calculations are carried out by computer, the results are, however, reproducible by anyone who uses the same numerical procedures. Factor analysis is often carried out in Q-mode: it seeks to find a limited number of samples which are "end-members," in the sense that other samples either resemble them closely, or can be made up by mixing two or more "end-members." Use of Q-mode generally involves computing with large matrices, and the number of factors retained may be larger than three, so that the results can only be (inadequately) visualized by plotting in two or three-dimensional factor space.

The subject goes well beyond the scope of this book, but luckily there is an entire volume, by Reyment and Jöreskog (1993), that is devoted to application of factor analysis methods in the earth sciences, and that provides a complete set of MATLAB functions for their implementation. Interesting examples of component and factor analysis are also presented in the book by Davis (1986).

6.5 Cluster Analysis

Cluster analysis, as generally understood, is a technique introduced by numerical taxonomists. It is best thought of as a way of illustrating graphically the relationships inherent in a correlation or similarity matrix. Its advantage is that it presents these relationships as an easily understood graphical tree (a *dendrogram*). Its disadvantage is that, for large data sets, some oversimplification of the relationships is inevitable. The full complexity of relationships among tens to hundreds of variables, samples, or taxa simply cannot be adequately expressed in a single two-dimensional diagram.

Cluster analysis proceeds by progressively reducing the similarity matrix to a smaller and smaller order. It can be applied to any kind of similarity matrix, in either R- or Q-mode. In what follows we assume for simplicity that the matrix is an R-mode correlation matrix. Reduction in size of the matrix is carried out by first seeking the two largest mutual correlations (i.e., if the variables are x_i and x_j we seek the pair that have a correlation r_{ij} such that r_{ij} is not only the largest correlation of the ith variable with any other variables, but also the largest correlation of the jth variable with any other variable). These two variables are then merged together and their correlation with all other variables is expressed by some method of "averaging" the correlations. The process is repeated until the correlation matrix is reduced to a single correlation coefficient, and the results are expressed graphically as a hierarchical tree.

The three crucial decisions in cluster analysis are (i) the choice of variables (and their scaling), (ii) the choice of the similarity measure, and (iii) the choice of the algorithm used to average the similarities, after merging two variables. Different choices give different results (e.g., see the examples shown in Davis, 1986, and Howard, 1991), and there are no real criteria to decide which choice is "right."

Programming cluster analysis for MATLAB proved rather challenging. We present one commonly used version: users can fairly readily modify it to use different algorithms for similarity averaging. Because of the complexity of the script, we show here only the basic script, not all the functions that it uses. These functions are included in the diskette distributed with the book.

```
function dendro5(R)
% dendro5(R)
% computes condensed forms S of the correlation matrix R
% until size(S) = [2,2], and plots dendrogram.  S is like a
% correlation matrix but has zeros in the principal diagonal.
% This version switches position of second most correlated
% object if it is not next to first. First pass determines
% switches necessary, R is then modified, and second pass
% draws dendrogram. Version for MATLAB 5
% Uses diagzero, maxcols, rswitch2, vswitch, doublev,
% shrinkm, oswitch, omerge, dendrox2
[r c] = size(R);        % determine size of R
v1 = [1:r];             % object matrix to record object positions
ro = ceil(r/2);
if ro < 10, ro = 10; end;
OK = zeros(ro, c);      % make object key matrix
OK(1,:) = v1;           % object indices in first row
for pass = 1:2
   k = r;
   S = diagnego(R);              % set -1's in diagonal (< min r)
   rl = 1.0;
   ru = 1.0;
   [c1, c2] = maxcols(S);        % determine cols with max correlation
   if (c2 - c1) > 1
```

```
      S = rswitch2(S, (c1+1), c2);  % reform correlation matrix
      if pass == 1
         OK = oswitch(OK, (c1+1), c2); % keep track of switches
      end
      c2 = c1+1;
   end
   if pass == 2                     % plot first part of dendrogram
      x = doublev([c1, c2]);
      y = [rl S(c1,c2) S(c2, c1) ru];
      plot(x,y);
      axis([0 r+1 -1 1]);           % and set up axes
      x3 = [0:r+1];
      set(gca,'XTick', x3(2:r+1), 'XTickLabel',v2);
         % for version 4 change to 'XTickLabels'
      title('Dendrogram');
      xlabel('Object number'), ylabel('Correlation, r');
      hold on;
   end % if pass == 2
   xx = [1:r];                      % x scale for plots
   rr = ones(1,r);                  % r scale for plots
   while k > 2                      % recalculate S until it is 2x2
      xx = dendrox2(xx, c1, c2);  % revise scales for plots
      rr(c1) = S(c1, c2);
      rr(c2) = [];
      SS = shrinkm(S, c1, c2); % merge cols c1 and c2 of S
      if pass == 1
         OK = omerge(OK, c1, c2); % and of OK on first pass
      end
      S = diagnego(SS);          % print out intermediate steps
      [c1,c2] = maxcols(S);
      if (c2 - c1) > 1            % switch cols if not adjacent
         S = rswitch2(S, (c1+1), c2);
         if pass == 1
            OK = oswitch(OK, (c1+1), c2);
         end
         c2 = c1+1;
      end
      if pass == 2
         x = doublev([xx(c1), xx(c2)]);
         y = [rr(c1) S(c1,c2) S(c2,c1) rr(c2)];
         plot(x,y);
      end
   k = k-1;
   end % while
```

```
    if pass == 1
        v2 = OK(:);    % v2 has elements of OK reordered as vector
        vi = find(v2); % find nozero elements
        v2 = v2(vi);   % redefine v2 using only nonzero elements
        R  = R(v2,v2); % use v2 to reorder R
    end
end % for
xl = (xx(c1)+xx(c2))/2;   % draw in last stem
x = [xl, xl];
rl = S(1,2);
y = [rl, -1.0];
plot(x,y);
hold off;
```

The function dendro5 begins by establishing an "object key" matrix OK, whose purpose is to keep track of all the merges and switches in position of the variables during the first pass through the program. Switches in position are necessary because the dendrogram cannot be drawn properly unless highly correlated variables are arranged next to each other in the tree. At each stage of the computation, each column of OK contains a list of the variables grouped together in the tree, as assembled up to that stage. At the end of the first pass, OK has just two columns. The elements are then reordered into a single column vector, which becomes the new order of variables across the top of the dendrogram.

Next the input matrix R has to be changed to S by using the function diagnego to substitute -1 for 1 in the principal diagonal. This has to be done because the variables to be merged are chosen by searching for the two columns in S that have the largest, equal correlation coefficients, using the function maxcols (at the end these may both be negative, but they are always larger than -1). After the columns c1 and c2 have been identified, c2 is switched with the adjacent column c1+1 if it is not already in that position (using rswitch2 for the S matrix and oswitch for the OK matrix. Merging of the S matrix is carried out using the function shrinkm, and of the OK matrix using the function omerge. In the second pass, the S matrix should need no further rearrangement by switching columns. A variety of graphics functions (doublev and dendrox2) are used to draw the tree at each stage of its computation.

The algorithm used to merge the two variables is to take the unweighted average of their correlations with other variables. For other possible choices, see Davis (1986, p.502–515), Dunn and Everitt (1982, p.77–85), or Howard (1991, p.175–196).

Most applications of cluster analysis are in taxonomy, of both modern and fossil organisms. Much of the literature is full of jargon related to that field: the book by Dunn and Everitt (1982) provides a brief, relatively jargon-free introduction. A classic application to the study of Bahamian sediments was given by Purdy (1963) and his data were further analyzed by Parks (1966). We used Purdy's data (in file purdyr.dat) to produce Figure 6.5. It shows that carbonates sediments on the Great Bahama Bank cluster into four (or five) groups, characterized by (i) coralline algae (1), corals (4), and *Halimeda* (2 – another type

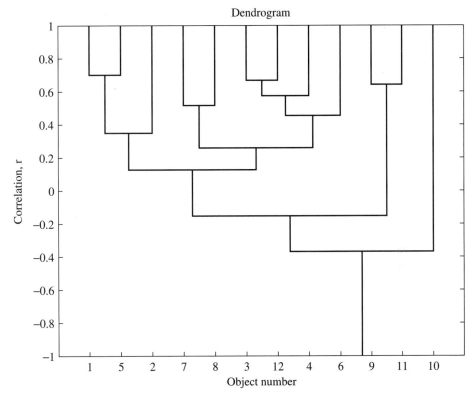

Figure 6.5   Dendrogram for constituents of Bahamian sediments, using data in Purdy (1963). In this example, the "Object numbers" are labels identifying various petrographic constituents.

of alga); (iia) fecal pellets (7), and mud aggregates (8); (iib) forams (3 and 4), mud (12), and molluscs (6); (iii) grapestones (9) and other cryptocrystalline grains (11); and (iv) oolites (10).

MATLAB may be used to produce small dendrograms, using up to about 30 variables. For larger dendrograms, a specialized clustering program should be used (see Afifi and Clark, 1990). A modern application in paleontology is given by Vasey and Bowes (1985).

6.6   Recommended Reading

Afifi, A.A. and V. Clark, 1990, Computer-Aided Multivariate Analysis. New York, Van Nostrand Reinhold, second edition, 505 p. (QA278.A33 A guide to using the major commercial statistics packages BMDP, SAS, and SPSS to perform multiple regression, factor and cluster analysis, and more. Emphasis on getting and interpreting results, rather than theory or computation.)

Anderson, T.W., 1984, An Introduction to Multivariate Statistical Analysis. New York, John Wiley and Sons, second edition, 675 p. (QA278.A516 A mathematically advanced text.)

Dunn, G. and B.S. Everitt, 1982, An Introduction to Mathematical Taxonomy. Cambridge University Press, 152 p. (QH83.D86 A concise introduction to classification techniques; accessible to those not deeply involved in biological taxonomy.)

Ferguson, J., 1994, Introduction to Linear Algebra in Geology. London, Chapman and Hall, 203 p. (A simple introduction, with many worked examples.)

Howard, P.J.A., 1991, An Introduction to Environmental Pattern Analysis. Park Ridge NJ, Parthenon, 254 p. (TD193.H69 Covers all the topics in this chapter, and more, with environmental examples.)

Le Maitre, R.W., 1982, Numerical Petrology: Statistical Interpretation of Geochemical Data. New York, Elsevier, (QE431.5.L33 Most of the book describes multivariate techniques including Principal Components, Factor Analysis, Multiple Discriminant Analysis, and Cluster Analysis.)

Parks, J.M., 1966, Cluster analysis applied to multivariate geologic problems. Journal of Geology, v.74, p.703–715 (Includes further discussion of the Bahamas sediment samples studied by Purdy.)

Potter, P.E., N.F. Shimp and J. Witters, 1963, Trace elements in marine and freshwater argillaceous sediments. Geochimica et Cosmochimica Acta, v.27, p.669–694 (One of the first applications in the earth sciences.)

Purdy, E.G., 1963, Recent calcium carbonate facies of the Great Bahama Bank: 1. Petrography and reaction groups. Journal of Geology, v.71, p.334–355. (Classic study, using cluster analysis.)

Reyment, R.A. and K.G. Jöreskog, 1993, Applied Factor Analysis in the Natural Sciences. Cambridge University Press, 371 p. (QA278.5.R49 A thorough treatment, with a complete set of MATLAB scripts.)

Rollinson, Hugh R., 1993, Using Geochemical Data: Evaluation, Presentation, Interpretation. Harlow, Essex (England), Longman, 352 p. (QE515.R75 Good examples of classification techniques applied to geochemical data.)

Chapter 7

Time Series

7.1 Introduction

Most data in the earth sciences are located in time or space or both. This fact is often ignored in the data analysis techniques that we have examined in earlier chapters. For example, Fisher's samples of iris plants were located in space: but he does not tell us their coordinates, because he knows (or assumes) that position in the colony did not affect their measurements. Generally, however, measurements do depend on when and where they are made: the discharge of a river varies with time (and so does the sediment content or chemical composition of the water). The oxygen isotope content of snowfall depends on the location, and also varies slowly over time because of the changing amount of light isotopes that are stored in the world's ice sheets. So a core of ice taken in Greenland shows progressive changes in isotopic composition with depth in the core, reflecting the age of the snowfalls that accumulated to form the different layers of ice sampled by the core.

In the earth sciences, stratigraphic position (depth in a core, elevation above datum in a measured section) is often a proxy for time, and measurements made at regular intervals across a section can often be analyzed using the same methods used to analyze true time series. Such measurements differ from random samples, because experience shows that samples taken very close together in either time or space tend to be very similar, no matter what is measured. Nature generally varies continuously from one time or place to the next, and sharp discontinuities are rare—though not unknown. For example, a succession of sandstones in a stratigraphic section may record a series of sudden discontinuities ("event beds") produced by storms or turbidity currents that interrupt the slow "background" sedimentation of fine-grained mud. But observation shows that there is generally a pattern of sandstone deposition that changes only slowly with time: "packets" of thick, coarse sand beds give way to "packets" of thinner, finer sand beds. Even sudden, erratic events may show changing patterns of occurrence with time.

In this chapter, we will examine techniques that have been devised to examine such changes. The data are commonly called *time series* even though they may actually be sequences of measurements made in space. The analysis of time series is a branch of data analysis that has developed largely independently of other numerical techniques, and has separate literatures in several different areas of application: the literature in signal processing has a jargon different from that in economic time series or turbulence theory. Luckily MATLAB has proved a particularly effective computing environment for analyzing time series, and has been adopted by many of those in the signal processing community: so a particularly rich set of numerical tools are available for application in other fields, once the barriers of disciplinary jargon are overcome.

7.2   Properties of Time Series

The following are properties of some, though by no means all, time series:

- Measurements are recorded at regular time intervals Δt. This is generally true in signal analysis, economic time series, and also in many physical and geophysical applications, such as meteorology (temperature, pressure, rainfall, etc.), hydrology (discharge of water and/or sediment), oceanography (output of wave staffs or buoys, and moored current meters), seismology (ground motions recorded by seismographs), and so on. Unfortunately, many geological time (or space) series consist of measurements that are irregularly spaced in time or space. Many of the time series recorded as electro-magnetic or geophysical signals are also measured at very short time intervals, and consist of many tens of thousands of highly precise measurements. Time series of interest to geologists generally consist of only a few hundreds of measurements: some outstandingly complete examples are measurements made on ice or deep-sea cores (about a thousand measurements, extending back about 250,000 years); daily measurements of temperature, etc., extending back some 300 years; the annual height of the Nile floodwaters extending back about 2500 years; the elevation of the Great Salt Lake, measured precisely every 15 days since 1847; the content of carbon dioxide in the atmosphere at Hawaii, measured every month since 1976.

- Time series often show both a gradual trend, and one or more periodic components. Time series whose properties change progressively with time are called *non-stationary*. Various degrees of *stationarity* are recognized by absence of trends in the mean value, the variance, or other statistical properties of the signal. The trend may be detected and removed from the data using the techniques of linear or polynomial regression already discussed in Chapter 5. Alternatively, taking the first difference ($d_i = x_{i+1} - x_i$) is often effective in removing a trend. This can be done easily using MATLAB's `diff` function. The following is a function that calculates and plots a first difference:

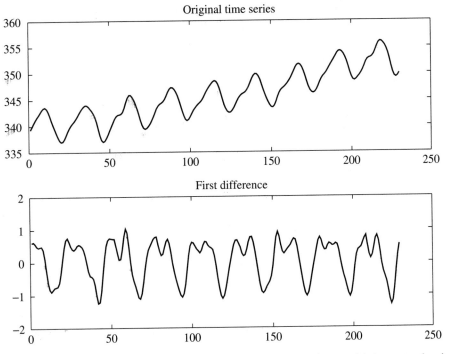

Figure 7.1 Carbon dioxide content of air (in ppm), measured every 14 days, starting in 1981, at Mauna Loa, Hawaii (upper plot); and the first difference (lower plot). Data from Garcia (1994).

```
function xd = fdiff(x)
% xd = fdiff(x)
% calculates and plots the first difference of the
% time series x.  Written by Gerry Middleton, Nov. 1996.
xd = diff(x);  % calculate differences
subplot(2,1,1);
plot(x,'LineWidth',1), title('Original time series');
subplot(2,1,2);
plot(xd,'LineWidth',1), title('First difference');
```

Figure 7.1 shows how effective this technique is in removing the trend from the well-known time series of carbon dioxide concentrations in air, measured at Mauna Loa (data from Garcia, 1994, in mauna.dat).

- Much of the effort in time series analysis is devoted to discovering the presence and significance of periodic components. Each periodic component is characterized by a particular period, or frequency (the inverse of the period). The *period* is measured as a time T: it is often (but perhaps unrealistically and unnecessarily) assumed that a time series has a *fundamental period* T_p, which is the period after which the whole series repeats itself. For a simple sinusoid the fundamental period is, of course, 2π. The *frequency* ν is simply $1/T$: if T is measured in seconds, then ν is measured in *hertz*. Often frequency is expressed instead as the *angular frequency* ω, equal to $2\pi\nu$ or $2\pi/T$. The fundamental angular frequency of a sinusoid is one. A periodic signal also has a *phase*, that is the part of the period through which the signal has advanced, measured from some arbitrary origin. A sine wave $x = \sin(t)$ that begins at $t = 0$ is considered to have zero phase, one that begins at $t = \phi$ is considered to have a phase of ϕ.

We will see in the next section that stationary time series may be characterized either in the *time domain*, by plotting the measured variable x as a function of time, or alternatively in the *frequency domain*, by plotting the magnitude and phase of its periodic components against their frequency. This is true even if the signal is not strictly periodic.

- The covariance or correlation of x at time t ($x(t)$) with x at some lagged time ($x(t+\tau)$, where τ consists of m time intervals Δt) is called the *autocovariance* or *autocorrelation*. Generally, the autocorrelation decreases from one at zero lag time to near zero at large lag times, but it may vary cyclically at intermediate times. This indicates that in most time series, x tends to be highly correlated with values of x measured a short time later, but uncorrelated with values measured a long time later. If the time series has a strong periodic component, this is indicated by a strong periodicity in the autocorrelation function, before it decreases to zero at very large time lags.

- All real time series also have a random component: this is generally called "noise" to distinguish it from the "signal" that interests the observer. This distinction is, however, almost entirely subjective: what constitutes "noise" for one observer may be a "signal" for another. Techniques to enhance the signal and suppress the noise constitute a large part of modern signal theory. Some time series, for example velocity measurements in turbulent flows, have a very large random component, and almost no truly periodic components. It is, nevertheless, useful to analyze them as though there were a large number of periodic components present.

7.3 Autocorrelation

In the statistical literature, the *autocovariance* is defined as the covariance between $x(t)$ and $x(t + \tau)$. For a time series consisting of N values of x, the covariance at a maximum lag of $\tau = m\Delta t$ is calculated on a maximum of $(N - m)$ values. A statistically unbiased estimate therefore divides the sum of products of deviations from the mean by the degrees of freedom $(N - m - 1)$. The unbiased estimate is, however, not always the one used in signal processing applications. The *autocorrelation* is defined as the autocovariance divided by the variance of x. A plot of the autocorrelation function is often called a *correlogram*, and the sample autocorrelation or autocovariance function is sometimes called the *serial correlation* or *serial covariance*.

Unfortunately, in the signal processing literature the autocorrelation is generally defined differently, as the sum of products of $x(t)$ and $x(t + \tau)$. In this case it is equal to the autocovariance plus the square of the mean value of x. In this book, we use the statistical definition.

Autocovariance and autocorrelation functions are readily calculated using MATLAB. The most efficient methods are indirect, and make use of MATLAB's fft function. The following function calculates the autocovariance and autocorrelation by a more direct, but less efficient method.

```
function r = scf(x, n, cor, bi)
% r = scf(x, n, cor, bi)
% if cor = 1, calculates the serial correlation function
% i.e., the sample autocorrelation function, of x, for
% up to n lags. n should be < N/4, where N is the length
% of x.  If cor = 0, the serial covariance function is
% calculated: bi = 1 means the biased estimate is
% calculated bi = 0 (the default) means the unbiased
% estimate is calculated.  Ref: articles by O.D.Anderson
% in the Encyclopedia of Statistics.
% Written by Gerry Middleton, November 1996
N = length(x);
xbar = mean(x);
xd = x - xbar;
ss = xd'*xd;
for i = 1:n+1
    xdlag = x(1:N+1-i) - xbar;
    if cor == 1
        r(i) = xd(i:N)'*xdlag / ss;
    elseif bi == 1
        r(i) = xd(i:N)'*xdlag / N;
    else
        r(i) = xd(i:N)'*xdlag / (N-n);
    end
end
lag = [0:n];
plot(lag, r,'linewidth',1);
if cor == 1
    title('Serial correlation function');
elseif bi == 1
    title('Serial covariance function (biased)');
else
    title('Serial covariance function (unbiased)')
end
```

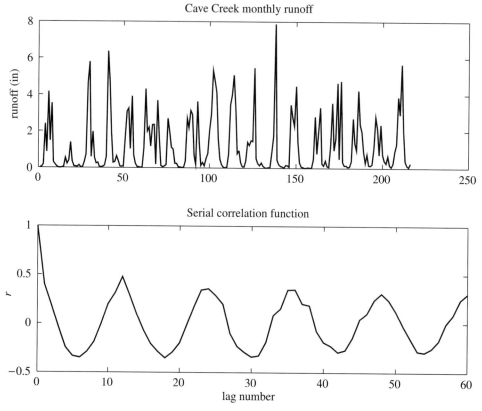

Figure 7.2   Upper diagram shows the time series of monthly runoffs, in inches, for Cave Creek, Kentucky. Lower diagram shows the autocorrelation function for the lags up to 60 months. Data from Haan (1977).

As an example, we use the time series of monthly discharges of Cave Creek, Kentucky, from October 1952 to September 1970 (data from Haan, 1977; reproduced in Davis, 1986, p.257; file `cvcrk.dat`). This stream shows a strong yearly maximum discharge, generally in March of each year. The strong yearly periodicity is reflected in the periodicity of the autocorrelation function: the peaks of the function show almost no decrease in height with increasing lag, showing that the yearly pattern of precipitation changes very little from one year to another over at least a 5 year (60 month) lag.

This type of autocorrelation function may be contrasted with that shown for most time series that lack such a strong periodic component. In Figure 7.3, we show the autocorrelation function for the assays of zinc in a vein, previously discussed in Chapter 1. It can be seen that there is positive correlation for lags up to 5 or 6 (distances of up to 12 meters apart in the vein), but no strong evidence for any periodicity. In other words, there are parts of the vein (of order 10 meters across) that are relatively richer or leaner than others in zinc, but there is no periodic, repeated enrichment in zinc along the vein. This corresponds well

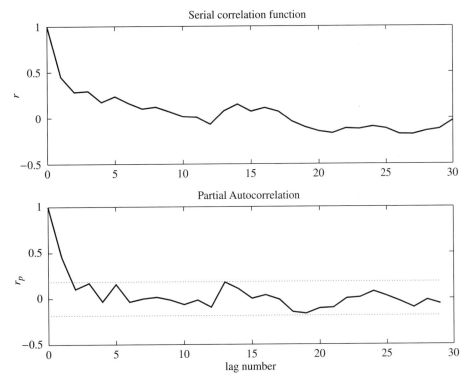

Figure 7.3   Upper diagram shows autocorrelation function for the De Wijs data for zinc in a vein (see Chapter 1). Lower diagram shows the partial autocorrelation function for the same data: the dotted lines indicate the confidence intervals for the partial correlations.

with our experience of how most natural properties are distributed spatially: there is local continuity, but regular spatial periodicity is rare.

A truly random time series would show an autocorrelation function that dropped suddenly from 1 at zero lag to near zero at lags equal to one or larger. Such autocorrelation functions are rare in natural phenomena, even for supposedly "chaotic" phenomena, such as turbulent flows.

As a final example, we show some unique data collected in a study of the Columbia River (file prdav.dat). At several sites in this large river, the U.S. Geological Survey measured velocity profiles using an array of ten current meters mounted on a vertical frame (Savini and Bodhaine, 1971). At one site, near Priest River Dam, velocities were recorded every minute for 66 minutes. Figure 7.4 shows that large variations in velocity were observed, even though the river discharge (controlled by the dam) was constant over the period of observation. These variations were undoubtedly produced by large scale turbulent eddies in the river (see Middleton and Southard, 1984, for further discussion). The autocorrelation function, plotted in the lower half of Figure 7.4 shows that the correlation falls off rapidly for lags up to 2 minutes, but rises again at lags of 3–4 minutes. The mean velocity was

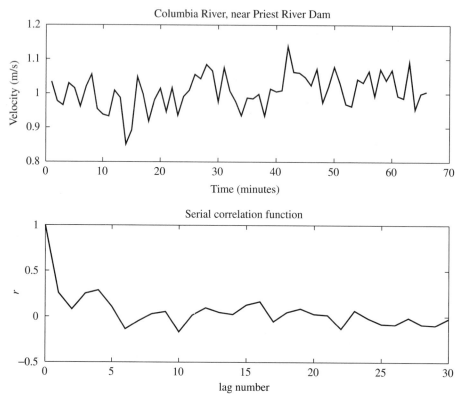

Figure 7.4   Upper diagram shows mean velocities measured in the Columbia River.
Velocities are averaged over the total depth (6.4 m). Lower diagram shows the
autocorrelation function for these measurements (data from Savini and Bodhaine, 1971).

about 1 ms^{-1}, so this suggests the possible influence of eddies with a horizontal scale of
about 200 m. This is a not-unreasonable hypothesis, as the width of the river was about 345
m, but the peak in the autocorrelation function is hardly large enough to provide convincing
proof of such a flow structure.

7.4   Partial Autocorrelation

A partial autocorrelation function consists of a series of partial autocorrelation coefficients,
which give the correlation between $x(t)$ and $x(t + k\Delta t)$, *after removing the effect of the*
terms from $x(t+\Delta t)$ *to* $x(t+(k-1)\Delta t)$. The square of each partial correlation coefficient,
therefore, gives the proportion of the sum of squares of deviations from the mean accounted
for by the kth term, after the effects of the preceding $(k-1)$ terms have been removed by
multiple linear regression (see Chapter 5).

Calculating and plotting the partial autocorrelation function can help the investigator decide what type of model can appropriately be applied to a time series. In many time series, the value at a time t seems to depend only on the value at the previous time $t - \Delta t$, or the previous two times $t - \Delta t$ and $t - 2\Delta t$. All the remaining variation can be accounted for by the addition of a random variable with constant variance. Such a model is called an *Autoregressive process*, or AR model, of first or second order. In theory, all partial autocorrelations beyond the first or second lag-times should be zero. In finite samples, this is not the case but a 95% confidence interval is given roughly by $\pm 2\sqrt{1/N}$ (Bras and Rodríguez-Iturbe, 1985, p.59).

The partial autocorrelation function can be computed from the autocorrelation function: it is a little complicated, because essentially what is required to compute k terms are k separate multiple regressions. One algorithm (probably not the most efficient) is given by Brockwell and Davis (1991, p.102) and Chatfield (1989, p.53–56). It has the advantage that it also yields the estimated coefficients for AR processes. The following MATLAB function is based on this algorithm:

```
function rp = pautoc(r, k, N)
% rp = pautoc(r, k, N)
% Calculates the partial autocorrelation for lags up to
% lag k from an autocorrelation series r, calculated from
% a time series of length N (e.g., r is from the output
% of function scf). k must be less than the length (number
% of lags) of r. Plots the result as a partial autocorrelation
% function. Note that rp(1) is just 1, and rp(2) is the
% autocorrelation at lag 1, i.e., the correlation of
% x(t) and x(t+1). Only after these two terms do the
% true partial correlations begin, e.g., rp(3) is the
% correlation between x(t) and x(t+2), holding x(t+1)
% constant, and so on. The algorithm is from
% Brockwell and Davis, 1991, Time Series: Theory and
% Methods, p.102. Written by Gerry Middleton, November 1996
rp = zeros(1,k);
rp(1) = 1.0;    % rp(1) is the (partial) correlation at lag 0
rp(2) = r(2);   % rp(2) is the (partial) correlation at lag 1
for j = 2:k     % these are the true partial correlations
   R = zeros(j);
   ri = r(1:j);
   R(1,:) = ri;
   for i = 2:j
      ri = [r(i) ri(1:j-1)];
      R(i,:) = ri;
   end
   phi = R\r(2:j+1)'; %*
   rp(j+1) = phi(j);
end;
```

```
lag = [0:k];
plot(lag, rp,'LineWidth',1), title('Partial Autocorrelation')
hold on;
y = 2*sqrt(1/N);
plot([0,k],[-y,-y],':',[0,k],[y,y],':');
hold off
```

To print out the AR coefficients omit the semicolon on the line marked with the asterisk (*).

The lower diagram in Figure 7.3 shows the computed partial correlation function for the De Wijs data. Only the first autocorrelation is significant—all the partial correlations at lags 2 or greater lie within the 95 % confidence intervals. We might conclude, therefore, that the zinc content in this vein can be modelled as a first order autoregressive process.

7.5   Fourier Series

Fourier series are sums of sine and cosine waves, of the type

$$f(t) = \frac{a_0}{2} + \sum [a_n \cos(2\pi nt) + b_n \sin(2\pi nt)] \tag{7.1}$$

where $n = 1, 2, 3, \ldots$. When $t = 1$, the sine and cosine waves repeat, because they are multiples of 2π— so in this notation, the "fundamental period," after which f(t) repeats itself exactly, is one. Fourier series can be written differently: with $\cos(nt)$ and $\sin(nt)$, the fundamental period is 2π; with $\cos(2\pi nt/T)$ and $\sin(2\pi nt/T)$ it is T. Unfortunately, there is no standard notation. Because there is always a fundamental period, it might be thought that Fourier series would only be useful to represent periodic functions. But in practice, they can be used to represent almost any time series as a sum of sine and cosine waves, no matter how angular and irregular that original time series seems to be—a result that even Fourier himself found "quite extraordinary" (Hubbard, 1996, p.9).

Note that Fourier series are not just any sums of sinusoids: the first two terms have period 1, the second two terms have period 1/2, the third two terms have period 1/3, and so on. One consequence of this is that the different pairs of terms are *orthogonal*, that is, the integral of their products is zero. This makes it relatively easy to fit a Fourier series to an observed time series (see Hubbard, 1996, for a simple discussion of the meaning and advantages of orthogonality).

Fitting a Fourier series to a time series consisting of N equally spaced data generally produces a sum of $N/2$ distinct pairs of sine and cosine terms, with different amplitudes a_n and b_n. The constant term a_0, can be made zero by first subtracting the mean value from each term in the time series. For large N the amount of computation required is large, but it can be reduced to the order $N \log N$ operations, by use of a technique called the *Fast Fourier Transform*. There is one disadvantage to using the Fast Fourier Transform—the number of input data must be a power of 2 (e.g., 64, 128, 256, 512, 1024,...).

MATLAB's implementation is the function `fft`, and is fully described in the Manual. Before describing how to use this function, however, we need to learn a little about an alternative way to represent Fourier series, using complex numbers.

7.6 Sinusoidal Functions in the Complex Plane

Complex numbers are those that have a real and an imaginary part, for example

$$z = x + iy \tag{7.2}$$

where x and y are real numbers and $i = \sqrt{-1}$. z is a complex number, with a real part x and an imaginary part iy. Complex numbers are useful for representing trigonometric functions because of *Euler's formula*

$$e^{i\theta} = \cos\theta + i\sin\theta \tag{7.3}$$

Figure 7.5 shows a line of unit length, making an angle of θ with the x axis. The ordinate is the imaginary axis iy. The x (real) component is therefore $x = \cos\theta$, and the y (imaginary) component is $iy = i\sin\theta$.

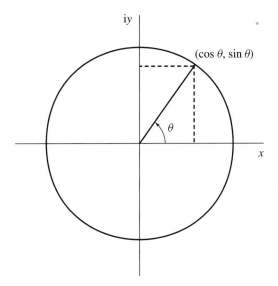

Figure 7.5   Representation of a complex number $z = x + iy$ in the complex plane.

Because of this relationship, the equation for a Fourier series may be written more simply, not as a sum of sine and cosine components but as a sum of a single complex exponential series

$$f(t) = \sum c_n e^{-2\pi i n t} \tag{7.4}$$

The complex coefficients c_n have real and imaginary parts a_n and ib_n. If we represent a single term of the Fourier sum in a complex plane diagram, then the length of the line would be its *amplitude*

$$|c_n| = \sqrt{a_n^2 + b_n^2} \tag{7.5}$$

and the angle that the line makes with the x-axis when $t = 0$ is its *phase* ϕ, given by

$$\cos \phi = \frac{a_n}{\sqrt{a_n^2 + b_n^2}}, \qquad \sin \phi = \frac{b_n}{\sqrt{a_n^2 + b_n^2}} \tag{7.6}$$

MATLAB's `fft` function determines the coefficients a_n and b_n for a time series x of length N. If N is a power of 2, then `fft` employs the fast Fourier algorithm; if N is not a power of 2 it employs a slower algorithm. See the Manual for details.

In practice, `fft` returns a two-column matrix. If there are N data, it consists of $N + 1$ rows. The first row has only one term, which is the real constant a_0 (so it should be equal to the sum of the series, because the sum over the series of all the sine and cosine terms is zero), and the next $N/2$ pairs of values are the coefficients of the cosine and sine terms corresponding to decreasing wavelengths or increasing frequencies. The central value (term number $N/2 + 1$) corresponds to the *Nyquist frequency*, after which the real coefficients are repeated symmetrically, and the imaginary coefficients are repeated antisymmetrically. For a numerical example, see the Manual.

We use the Columbia River data to illustrate the use of MATLAB's `fft`. The first 64 minutes of data were analyzed, using the deviation from the mean value as input. Figure 7.6 shows a stem-plot of the coefficients a_n and b_n, produced by the following script:

```
load prdav.dat;
x = prdav - mean(prdav);
x = x*0.305;       % convert from fps to m/s
Y = fft(x, 64);
n = [0:63];
subplot(2,1,1);
stem(n, real(Y)), title('a(n)');
subplot(2,1,2);
stem(n, imag(Y)), title('b(n)');
```

Note that the data were first standardized by subtracting the mean: it is not necessary to do this, but the plot of a_n would be distorted by the fact that a_0 for the original data is much larger than any other term (a_0 is equal to the mean value, times the number of data). Note also that the a_n values are symmetrical about $n = 32$, and the b_n values are antisymmetrical about $n = 32$ (equal in magnitude, but reversed in sign). Instead of using the series index n, we could have used a frequency scale for the abscissa, prepared using the MATLAB command `f = (0:63)/64`.

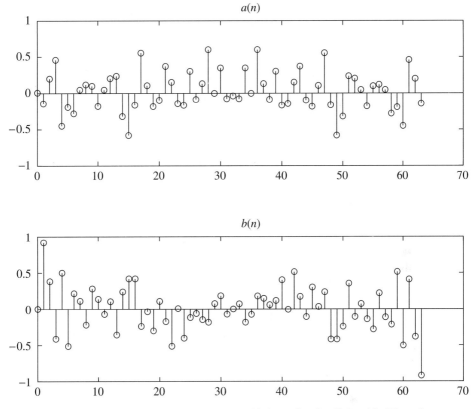

Figure 7.6   Stem plot of the Fourier coefficients for the Columbia River data

This illustrates an important fact about Fourier series: only the terms up to $n = N/2$ contain any useful information. The frequency corresponding to this midpoint (0.5 per minute) is called the *Nyquist frequency*. The significance of the Nyquist frequency is simple: the next higher frequency in our example would be one per minute, but since that is also the frequency of sampling it is clear that we cannot have any knowledge of velocity fluctuations taking place at that (or higher) frequencies—we simply do not have samples sufficiently closely spaced to detect periodic variations taking place on such a short time scale. The sampling theorem due to Nyquist (see Hubbard, 1986, p.221–223 for a proof) tells us, however, that we *can* reconstruct all frequencies in a signal less than ν, by sampling at a frequency of 2ν. Waves with a frequency higher than the Nyquist frequency appear to the observer, who only "sees" them at intervals larger than their period, to be much longer than they really are—a phenomenon (known as *aliasing*) that is illustrated visually by the slowly rotating spokes of wagon wheels observed in movies shot at only 24 frames per second.

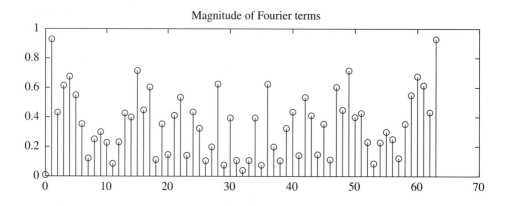

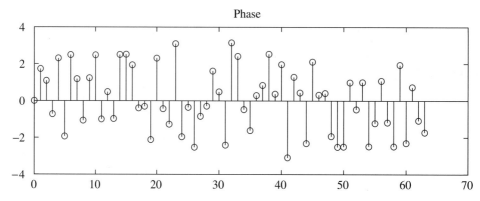

Figure 7.7   Stem plot of the Fourier coefficients for Magnitude and Phase of the Fourier
terms for the Columbia River data

An alternative, and more meaningful, way of representing the results of Fourier analysis
is by plotting the results as the magnitude and phase of the series terms. In MATLAB, this
can be done using the following script:

```
load prdav.dat;
x = prdav - mean(prdav);
x = x*0.305;        % convert from fps to m/s
Y = fft(x, 64);
n = [0:63];
subplot(2,1,1);
stem(n, abs(Y)), title('Magnitude of Fourier terms');
subplot(2,1,2);
stem(n, angle(Y)), title('Phase');
```

The result for the Columbia River data is shown in Figure 7.7. Once again, note that the magnitudes are symmetrical about $n = 32$ and the phases are antisymmetrical about this point. In some applications, it can be seen that the magnitudes of terms decrease dramatically at higher frequencies, so that these terms can be neglected. This is not the case for the Columbia River data.

Note that the result of Fourier analysis is to *transform* the data from a time series to a series of frequencies. For this reason, it is common to speak of the function of frequencies $F(\omega)$ as constituting a *Fourier transform* of the original time series $f(t)$. ω is the angular frequency defined as $\omega = 2\pi\nu$. For discrete time series, the frequency ν also takes discrete values $\nu = n/NT$, $\quad n = 1, 2, \ldots, N$, where T is the total duration (in time) of the time series, and N is the total number of time intervals in the sample. The Fourier transform is given by

$$F(\omega) = \sum f(t)e^{-i\omega t} \tag{7.7}$$

This function contains all the information present in the original time series. The original time series can be restored exactly (except for numerical rounding errors) by taking the *inverse Fourier transform*

$$f(t) = \frac{1}{2\pi} \sum F(\omega)e^{i\omega t} \tag{7.8}$$

Note that these two transforms are defined in slightly different ways by different authors. The important point is not the definition used, but the fact that it is possible to express the same information in two different ways: as a time or frequency series. In signal processing, these two contexts in which the data may be viewed or processed are called the *time domain* and the *frequency domain*.

In the applications described in this book, the time series consists of real numbers, and the frequency series consists of complex numbers, but this is not always the case (MATLAB allows for the possibility that the time series is complex). We have also seen that the complex frequency series may be regarded instead as being two series of cosine and sine components, or alternatively, two series of amplitudes and phases.

7.7 Spectral Analysis

One way of representing the data in the frequency domain is to plot the contribution which each frequency makes to the total variance of the time series. This representation has been given various names: the *variance spectrum* (in statistics), the *power spectrum* (originally in electrical engineering), or the *spectral density function*. The result of performing spectral analysis on a finite, discrete time series is to produce a finite, discrete spectrum, showing the estimated variance (amplitude squared) of each frequency in the series, up to the Nyquist frequency. This is called the *periodogram* (see Chatfield, 1989, p.110–111, for a discussion of the different ways this term has been defined by other authors). In MATLAB it can be produced in two different ways. The time series should first be standardized by subtracting the mean value (otherwise the first term, a_0^2, may distort the plot, unless a logarithmic scale is used). The first method (given in the MATLAB manual) is implemented by the following function:

```
function v = pdg(x, len, DelT, unit)
% v = pdg(x, len, DelT, unit)
% Function to calculate the periodogram (unsmoothed
% variance spectrum) of a time series x(t)
% len is the length to be analyzed -- it can be equal to
% or less than the length of x (N), and need not be a power of 2;
% or it can be the next power of 2 larger than N.
% DelT is the sampling interval, and unit is
% the (time) unit in which it is measured, e.g., 'minute')
% Written by Gerry Middleton, November 1996
N = length(x);
N2 = floor(N/2);
xbar = mean(x);
s2 = cov(x);
g = 1 - exp((log(0.95) - log(N2))/(N2-1));
x = x - xbar;
Yn = fft(x, len);
len2 = len/2;
Pyy = Yn .* conj(Yn)/(len*len);
f = (0:len2)/(len2*2*DelT);
plot(f,2*Pyy(1:len2+1),'LineWidth',1);
title('Periodogram -- arithmetic variance scale');
xlabel(['Frequency (per ',unit,' )']);
ylabel('Variance');
hold on;
plot([f(1) f(len2)], [2*s2*g 2*s2*g],':');
hold off;
figure;
semilogy(f(2:len2+1),2*Pyy(2:len2+1),'LineWidth',1);
title('Periodogram -- logarithmic variance scale');
xlabel(['Frequency (per ',unit,' )']);
ylabel('Variance');
hold on;
plot([f(1) f(len2)], [2*s2*g 2*s2*g],':');
hold off;
v = Pyy;
```

The essence of this function is the determination of the fast Fourier transform by Yn = fft(x,len) and the determination of the variance (power) components by Pyy = Yn .* conj(Yn) / (len*len). The rest is just a matter of proper scaling and plotting. In calculating the Fourier transform, we make use of the version of fft that permits truncation of the input series to a length `len` or expansion of it to that length by "padding with zeros." Padding with zeros is a computational technique which enables us to use a complete time series, even when its length N is not a power of 2. So, for example, if $N = 1000$, we set `len` equal to 1024, rather than 512. Whatever we do, the Nyquist frequency remains the same: it is $1/2 * \Delta t$. So we are interested in plotting only the first half of Pyy, i.e., the part up to the Nyquist frequency, or the first $(L/2) + 1$ terms, where L is the length specified by `len`. The frequency scale `f` goes from zero (for the constant term) up to the Nyquist frequency (in plotting variance on a logarithmic scale, we must omit the first term, which was set to zero by standardizing the data to zero mean). In calculating Pyy we divide by the length of the series squared, not just by the length as recommended in the MATLAB manual. The reason is that we want the values in Pyy to sum to the total variance. Also, if we plot only the first half of Pyy then we must multiply the values by 2 to obtain the correct contributions of each frequency to the total variance. The horizontal dotted line indicates the variance that must be exceeded by the largest peak, for that peak to be considered significant at the 95% confidence level (computed from a formula given by Davis, 1986, p.258).

Figure 7.8 (upper diagram) shows the result of applying this function to the Columbia River data. None of the peaks shown can be considered statistically significant. One of the largest peaks is at a frequency of 0.25 per minute, corresponding to the wavelength peak of 4 minutes detected in the correlogram. Unfortunately, this time series is not long enough to provide a good test of the hypothesis that there are large eddies whose horizontal scale is determined by the width of the river.

An alternative way to calculate the variances is to use the square of the amplitude of the Fourier transform, i.e., by substituting

```
Pyy = (abs(Yn)) .^ 2 /(len*len)
```

in the function given above. The results are identical. Note that not all the information contained in the original time series is present in the spectrum: all the information about phases is lost. It can be shown, however, that the autocovariance function and the variance spectrum are a Fourier transform pair (as the time series and the Fourier series are). The periodogram is therefore the frequency domain representation of the information displayed in the time domain by the autocovariance function. This means that the autocovariance function can also be computed by first using the `fft` function to obtain the spectrum Pyy and then using the inverse Fourier transform `ifft` to obtain the autocovariance function from Pyy. Though this may seem a roundabout method, it generally computes faster than the direct method presented earlier in this Chapter.

Another interesting application of the Fourier transform is to generate simulated time series, different from the original series, but which have the same variance spectrum as the original (therefore their statistical properties are essentially identical). This can be done using the following function:

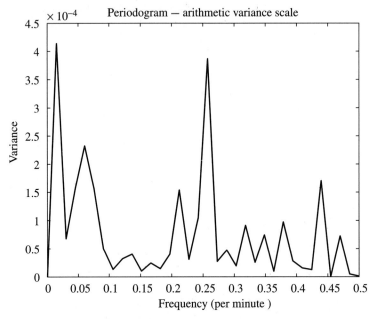

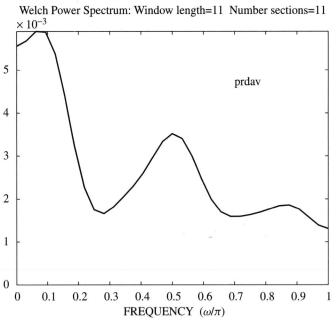

Figure 7.8   Upper diagram shows periodogram for Columbia River data; lower diagram
shows smoothed version using Welch's method (see text).

```
function xs = surrog(x)
% xs = surrog(x)
% Generates a surrogate time series having the
% same variance spectrum as the original series x
% The surrogate transform has randomized phases.
% Plots the two series for comparison.
% Written by Gerry Middleton, November 1996
N = length(x);
Y = fft(x);
mag = abs(Y);    % determine the amplitudes
phi = angle(Y); % and phases
% now generate random nos between -pi
% and pi, arranged antisymmetrically
phi1 = 2*pi*(rand(1,N/2) - 0.5);
phi2 = -phi1(N/2:-1:1);
% combine with the first term of phi to
% give a random set of phases
rphi = [phi(1) phi1 phi2(2:N/2)]';
% and combine with the original amplitudes
S = mag .* exp(i*rphi);
% finally generate xs, and plots
xs = real(ifft(S));
subplot(2,1,1);
plot(x,'LineWidth',1), title('Original data');
subplot(2,1,2);
plot(xs,'LineWidth',1), title('Surrogate data');
```

Figure 7.9 shows one result of applying this function to the original Columbia River data. Repeated applications, of course, produce different surrogate series.

The periodogram can be interpreted as a sample estimate of the *spectral density function* of the time series "population." That is, if the time series (assumed stationary) were of infinite length, or alternatively if we had an infinite number of realizations of it, over a given time period, then we should be able to determine the density function that gives the variance to be attributed to all frequencies between ν_1 and ν_2. Generally, this is what we want to know: though our data are discrete and finite in number, the real phenomenon is generally continuous, and conceptually (at least) an infinite number of samples could be drawn from this continuum.

It turns out that the periodogram has extremely bad sampling characteristics: the variance of the estimate for any particular frequency does *not* decrease with increasing sample size N (i.e., it is *not* a consistent estimator of the spectral density function)! If the time series is a purely random series with total variance s^2, then it can be shown (Chatfield, 1989, p.113) that the contribution estimated for any particular frequency has standard deviation s^2/π. Plotting twice this value on the periodogram gives some indication of whether or not any peaks are likely to be significantly different from zero. In the function pdg,

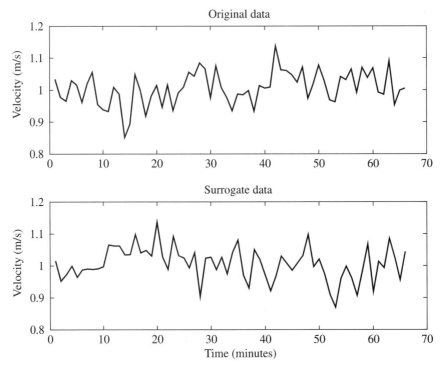

Figure 7.9   Upper diagram shows original time series (velocities in feet per second measured near Priest River Dam); lower diagram shows simulated (or "surrogate") time series that has the same variance spectrum, but randomized phases.

however, the line plotted is a somewhat smaller interval, using a criterion given by Davis (1986, p.258) for the significance of the largest peak only. Hegge and Masselink (1996) give further discussion of confidence limits for spectral estimates, using the Chi Square distribution, and provide a complete set of MATLAB functions for their implementation.

    Improved estimators of the spectral density can be obtained by various filtering or smoothing techniques. A good introduction is given by Diggle (1990), and a full discussion is given by texts such as Hayes (1996), Kay (1988), Marple (1987) and Percival and Walden (1993). The methods go well beyond the scope of this text. One simple technique (called Welch's method, or "weighted overlapping segment averaging," or WOSA) consists of dividing the original times series into a number of overlapping segments, determining the periodogram of each segment, and then averaging the result (for details see Percival and Walden, 1993, p.289–294). It is implemented in the MATLAB function Welch2, which is included (by permission of the authors) in the diskette distributed with this book and in the Student Edition of MATLAB 5 by the function spectrum. Figure 7.8 shows this technique applied to the Columbia River data. The frequency scale is set so that the Nyquist frequency is equal to 1. The smoothing shows two major peaks, one at very low frequencies (corresponding to wavelengths of the order of 1000 m), the other (as before) at about half the Nyquist frequency (corresponding to wavelengths of about 200–250 m). Smoothing is certainly effective, even for this very short time series, but as expected, is accompanied by a loss of resolution. For MATLAB files implementing the Welch and other averaging and smoothing techniques see Hayes (1996, Chapter 8).

Spectral analysis of unevenly spaced data is discussed by Press et al. (1989 : see the second FORTRAN edition).

7.8 Filters

A filter can be defined as any process that modifies a signal. In the signal processing literature, filtering generally refers to modification in the frequency domain, e.g., removal or suppression of certain unwanted frequencies, which may be regarded as suppression of "noise" and enhancement of the "signal." Filtering may be carried out by analog methods, e.g., by the use of mechanical or electrical systems: a shock absorber, or an electrical delay circuit, is a type of filter. We are concerned here only with numerical methods. Before the development of the fast Fourier transform and fast computers (i.e., before 1965) most numerical filtering was carried out in the time domain, where it is generally described as averaging or smoothing the signal. We have already seen a few examples of these techniques, and will describe a few more before going on to discuss (briefly) some aspects of filtering in the frequency domain.

7.8.1 Moving Averages and Convolution

First, consider what is involved in taking a *moving average* of a time series. A simple 3-point moving average (called a "running mean" in signal processing) is obtained by averaging the observed value of x at time t, with the values immediately preceding and following that time:

$$g(t) = (f(t - \tau) + f(t) + f(t + \tau))/3 \tag{7.9}$$

which may also be written

$$g(i) = \sum_{j=-1}^{j=1} w_j f(i + j) \tag{7.10}$$

In this second form, we have allowed for the possibility that the weights w_i might not all be equal (to 1/3) as in the first form of the equation. Obviously, this is a type of filter, because it transforms the original signal $f(i)$ into a different form $g(i)$.

This type of filtering can be described by a vector operation called *convolution* and generally denoted by a star, * (e.g., Davis, 1986, p.114-116). Two vectors x and y have the convolution z where $z = x * y$ when

$$z_i = \sum_{j=0}^{j=i} x_j y_{i-j} \tag{7.11}$$

In this representation, filters are just special vectors designed to be combined by convolution with data vectors. We can perform a three-point moving average by the convolution

$$z = x * f/3, \quad f = [1\ 1\ 1]'$$

MATLAB implements this using the command z = conv(f,x)/3. As an example, consider a moving average of the series

$$a = [1\ 3\ 5\ 7\ 2]$$

given by Davis (1986, p.116). The result using MATLAB is

$$1/3\ 4/3\ 3\ 5\ 14/3\ 3\ 2/3$$

The third term is obtained by averaging the first three terms of a, the fourth by averaging terms 2–4, and the fifth by averaging terms 3–5. The first term averages 1 with two zeros, the second averages 1 and 3 with zero, and similarly for the last two terms in the smoothed time series. Of course, these terms at the beginning and end of the smoothed series are not really valid averages as we have "padded" the time series with leading and trailing zeros in order to calculate the averages (MATLAB actually uses fft and ifft to calculate convolutions in the frequency domain).

7.8.2   Differencing

In an earlier section, we discussed the use of differencing to remove a trend in a time series. Obviously, this technique can be regarded as a form of filtering in the time domain. It can also be represented as a convolution of the time series and the vector b = [1 -1]. The result of applying MATLAB's conv operator to the vector a used in the previous example is

$$c = \text{conv}(a, b) = [1\ 2\ 2\ 2\ -5\ -2]$$

Once again, convolution adds one leading and one trailing term to the differenced series.

Second order differencing, i.e., $h(t) = f(t + 2\tau) - f(t)$, may be carried out by convolution of $f(t)$ with $g = [1\ 0\ -1]$.

7.8.3   Removing Seasonal Components

A simple way to remove a seasonal trend, for example from monthly observations, is to determine the long term monthly averages, and subtract them from the data. Note that this assumes that the time series is stationary (except for seasonality), that is, there is no trend in the mean that, in turn, determines the size of the seasonal effect. For example, the Hawaiian carbon dioxide data presented in Figure 7.1 shows a strong trend, which can be removed by first differencing. It is reasonable to assume, however, that the seasonal variation of carbon dioxide (which is also a prominent feature of the series) is proportional to the mean content of carbon dioxide in the air, rather than being independent of it. This effect is barely perceptible in the data, because the mean value of carbon dioxide has varied relatively little over the period of observation. These complications are discussed further by Chatfield (1989, p.17–18).

The following MATLAB function removes a seasonal trend by subtracting the seasonal average:

```
function xd = deseas(x, m)
% xd = deseas(x, m)
% Function to remove seasonal component from a
% time series x, by subtracting the mean seasonal
% values.  m is the number of terms in a season,
% e.g., for a yearly season, and monthly data m=12.
% The length of x should be an integral multiple of
% m, i.e., N = m*n.
% Written by Gerry Middleton, November 1996
N = length(x);
n = N/m;
X = reshape(x, m, n);
X = X';
mbar = mean(X);
X = X - ones(n,1)*mbar;
xd = reshape(X', N, 1);
subplot(2,1,1);
plot(x, 'LineWidth',1);
title('Original data');
subplot(2,1,2);
plot(xd, 'LineWidth', 1);
title('Seasonally adjusted data');
```

Figure 7.10 shows the result of applying this function to the Cave Creek runoff data. Calculating the autocorrelation or variance spectrum of this seasonally adjusted data shows that the seasonal component has been effectively removed by this technique—but visual inspection of the figure suggests otherwise. It is true that there is no longer any yearly peak, but there remains an irregular alternation of maxima and minima spaced about one year apart. In other words, removing the seasonal mean does not remove the large seasonal variability: there is more variation in the annual flood than in the base flow of this stream. This large seasonal variability is, however, not reflected in the variance spectrum, which shows a minimum at frequencies of one per year. If the seasonally adjusted data is further transformed by taking absolute values, then both the autocorrelation function and variance spectrum once more show a strong yearly periodicity. The moral is that one should look hard at data both before and after applying filters.

7.8.4 AR, MA, FIR and IIR

The dependence of $f(t)$ on earlier values of the series $f(t - \tau), f(t - 2\tau), \ldots$ is revealed by the autocorrelation and partial autocorrelation functions. It can also be made the basis of an *autoregression (AR) model* of the time series. For example, suppose we represent the

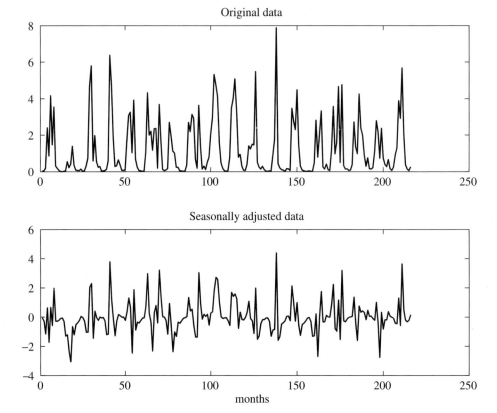

Figure 7.10   Cave Creek data, seasonally adjusted by removing the monthly averages.

time series by $x(n)$, where $n = 1, 2, \ldots$ (so $x(1) = x(t)$, $x(2) = x(t + \tau)$ and so on), then the regression equation for a second order AR process is

$$x(n) = \alpha_1 x(n-1) + \alpha_2 x(n-2) + \epsilon(n) \tag{7.12}$$

$\epsilon(n)$ is the residual, which in the model is assumed to be a random process with zero mean and variance σ_ϵ^2. Note that this equation can be regarded as a type of filter, which can produce $x(n)$ by taking a weighted average of the two previous terms in the time series. The random variable is an "error" term that is introduced to take account of (random) differences between the real time series and the predictions of the model. Some AR filters, when applied in the frequency domain, are called *infinite impulse response (IIR) filters*, because the effect of any change in $x(n)$ (the "impulse") is felt for an infinite number of time intervals in the future, though the effect decreases exponentially with time. One might think that because $x(n)$ depends only on the values for the previous two time intervals, that the effect of an impulse also persists for only two time intervals, but this is not true since the value at $x(n-1)$ was determined by the value of the two previous time intervals, and so on *ad infinitum*.

Another possible model for a time series is the *moving average (MA) model*. An example of a second order MA model is

$$x(n) = \beta_0 z(n) + \beta_1 z(n-1) + \beta_2 z(n-2) \tag{7.13}$$

From our earlier discussion of moving averages, we see that this model corresponds to generating a time series $x(n)$ by taking a three-point moving average of a random process $z(n)$. Therefore, it, too, can be regarded as a type of filter. When this type of filter is applied in the frequency domain it is called a *finite impulse response (FIR) filter*, because it depends only on three values of the random variable z, each of which are independent of all other values. Note that the random variables, in MA processes, are an integral part of the model: they are not merely introduced to take account of discrepancies between the predictions of the model and the real time series, as was the random element of the AR model.

Models that combine AR and MA are called ARMA models: their use is popular in the analysis of economic time series, and has been expounded at length by Box and Jenkins (1994; see Gottman, 1981, and Pankratz, 1983, for less mathematical treatments). For applications of ARMA models in the earth sciences see Bennett (1979) and Bras and Rodríguez-Iturbe (1985). Some MATLAB routines for their implementation are available in the Signal Processing and System Identification Toolboxes produced by The MathWorks, Inc. A set of routines written by Alois Schloegl is also available (as `tsa.zip`) on the Web, at

`http://www-dpmi.tu-graz.ac.at/~schoegl/matlab`

A simple example of a process that combines both AR and MA elements is a *random walk*. A one-dimensional random walk $x(n)$ is generated by taking the cumulative sum of a series of random numbers $z(n)$. Its value at the nth step is therefore given by

$$
\begin{align}
x(n) &= x(n-1) + z(n) \tag{7.14} \\
&= \sum_{i=1}^{n} z(i) \tag{7.15}
\end{align}
$$

The first term on the right hand side is an autoregressive term, and the second term is a moving average term. This should not be regarded as simply a first order AR model, because $z(n)$ is incorporated into the model itself. Nor should it be regarded as simply a MA process, because to do so we would have to change the range of the averaging at each step. However, it can be shown (e.g., Kay, 1988, p.112) that any AR process can be produced by an infinite order MA model, and any MA process can be produced by an infinite order AR model, so these distinctions do not have absolute validity.

Random walks are often used as models of natural time series: they are non-stationary and have fractal characteristics. We will return to them in Chapter 10.

7.8.5 Filtering in the Frequency Domain

Suppose we want to remove "noise" from a time series. By "noise" we generally mean high frequency measurement error, and natural variation, which makes it difficult to observe the lower frequency "signals" that are of interest. We could try to do this by using a moving average, but it seems clear that the proper place to filter out high frequency noise is in the frequency rather than the time domain. What we require is what the signal processing community calls a *low-pass filter*.

We can formalize much of the discussion of filters given earlier by writing a *standard difference equation* that specifies the way a signal $x(n)$ is changed into a filtered signal $y(n)$:

$$y(n) = \sum_k b(k)x(n-k) - \sum_k a(k)y(n-k) \tag{7.16}$$

where the summation extends up to some *order m*. Note that this is essentially a combination of the AR and MA models (i.e., an ARMA model) if we assume that the $x(n)$ are random variables (think of Equation 7.16 as a combination of Equations 7.12 and 7.13). In the time domain, the effect of the filter is specified by the two sets of coefficients $a(k)$ and $b(k)$. We can move from the time to the frequency domain by use of an appropriate transform, in this case called the z-transform. Then the effect of the filter is specified using a *transfer function* $H(z)$, where $z = e^{i\omega t}$ is a complex number. The relation between the two domains (not derived here) is given by the equation

$$
\begin{aligned}
H(z) &= \frac{B(z)}{A(z)} \tag{7.17} \\
&= \frac{b(0) + b(1)z^{-1} + b(2)z^{-2} + \ldots + b(m)z^{-m}}{a(0) + a(1)z^{-1} + a(2)z^{-2} + \ldots + a(m)z^{-m}} \tag{7.18}
\end{aligned}
$$

If the $a(k)$ are all zero, except for $a(0) = 1$, then the filter is a FIR filter, otherwise it is an IIR filter.

MATLAB uses these definitions in the implementation of its `filter` function (this function, and some others described in this section are available in the Student Version of MATLAB: they are part of the Signal Processing Toolbox, and are not included in the Professional Version). For example, we can implement a first order difference by the following MATLAB commands:

```
x = [1  3  5  7  2];
a = 1;
b = [1 -1];
[y, state] = filter(b,a,x);
```

y contains the filtered signal, and `state` contains the final set of values used by the filter (which may be used to continue filtering another segment of the signal—see the Student Manual). In the example, y = [1 2 2 2 -5] and `state` = -2 (compare with Section 7.8.2). For a three point moving average, change b to [1/3 1/3 1/3], and compare with Section 7.8.1).

The real value of this approach is that it allows the user of MATLAB to create filters with well-known properties, such as the Butterworth, Chebyshev, and Yule-Walker filters, adapted to the particular task required. Full details cannot be given in this text: an elementary treatment is found in Etter (1993, Chapter 13), a concise review is given by Chatfield (1989, Chapter 9), and more advanced treatment in a text on signal processing such as Oppenheim and Willsky (1983). We confine ourselves to demonstrating how MATLAB may be used to design a low-pass Butterworth filter and apply it to smoothing a time series.

One of the problems in filter design is to produce a filter that removes some frequencies, but passes others in an essentially undistorted form. The Butterworth filter has characteristics that make it a good choice as a low-pass filter. It is implemented in MATLAB as the function `butter`. The input parameters are m, the order of the filter (m+1 is the number of coefficients a(k) and b(k)), and the normalized cutoff frequency Wn. This is the upper limit of frequency that will be passed by the filter, expressed as a proportion of the Nyquist frequency. The output consists of the matrices of (complex) coefficients B and A. Increasing the number of terms used increases the sharpness of the cutoff, but causes loss of information at the beginning and end of the filtered series.

The following sequence of MATLAB commands applies a Butterworth filter to the Cave Creek time series. Recall that the Nyquist frequency for this monthly time series was 0.5 per month. The annual periodicity had a frequency of $1/12 = 0.083$ on this scale, which would be $1/6 = 0.167$ on the normalized scale required for input to `butter`. We use only 10 coefficients, which does not produce a very sharp cutoff, so choose a value of 0.3 for the cutoff frequency. It is assumed that `cvcrk.dat` has already been loaded into memory: then the application of the filter requires only the following two commands:

```
[B, A] = butter(10, 0.3);
[y, state] = filter(B, A, cvcrk);
```

The result is plotted in Figure 7.11, below the plot of the original, unfiltered time series. Note that there is some displacement of the time series to the right, and some loss of information at the upper end (for a more realistic plot one could rescale the abscissa to allow for this effect). The yearly peaks can be seen more clearly: the height of the peaks has been reduced, and the relative heights changed in some cases. For example, the highest peak flow, at about 140 months on the original time scale, has been reduced from about 800 original units (hundredths of an inch of runoff) to about 250 units after smoothing, and it is no longer the highest peak. These effects are due to the smoothing process, which reduces the height of sharp peaks, like the one at 140 months, much more than it reduces the height of broad peaks, like the one at 105 months. In this respect, the Butterworth (or any other frequency domain) filter is not much different from a moving average applied in the time domain. The ease of application, however, allows one to experiment with different designs of filter, to obtain the effect desired. The characteristics of the filter specified can be examined using MATLAB's `freqz` function (see the Manual, and Etter, 1993, p.333–337 for details).

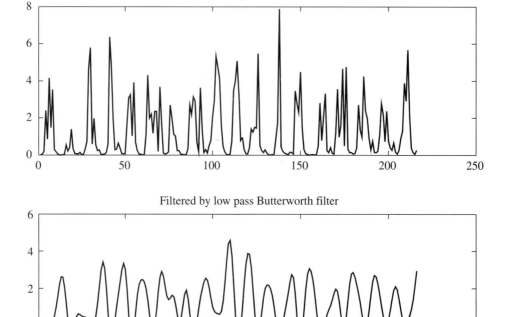

Figure 7.11   Effect of a low-pass Butterworth filter on the Cave Creek time series. Upper diagram shows original time series; lower diagram shows effect of filtering with a 10 term filter, with the cutoff set at 0.3 times the Nyquist frequency.

Figure 7.12 shows a final example of smoothing using a Butterworth filter. The bedform profile is the same one used to illustrate the use of splines in Chapter 5: as before it is plotted with a vertical exaggeration of 4. The original profiles showed some small irregularities: these could be real, or due to measurement errors. A Butterworth low-pass filter (with five terms and a cutoff of 0.3) removes them rather effectively, while still preserving most features of the bedform shape; but notice the distortion at the beginning of the profile.

7.9   Recommended Reading

The following are informal introductions to Fourier series and other topics in signal processing:

Hubbard, B.B., 1996, The World According to Wavelets: The Story of a Mathematical Technique in the Making. Wellesley MA, A.K. Peters, 264 p.(QA403.3H83)

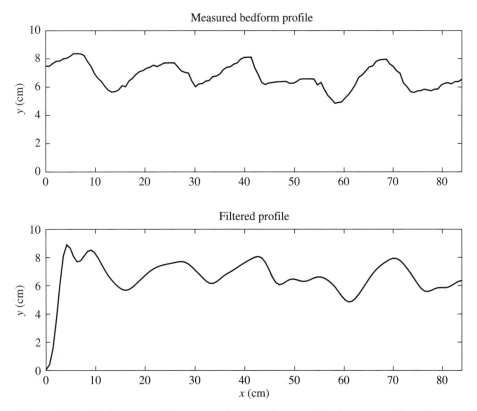

Figure 7.12   Bedform profiles: upper diagram shows original measured data; lower
diagram shows profile smoothed using a Butterworth filter.

Steiglitz, K., 1996, A Digital Signal Processing Primer: with Applications to Digital Audio
and Computer Music. Menlo Park CA, Addison-Wesley, 314 p. (TK5102.9.S74)

A large number of books describe classical time series analysis. The simplest introductions are given by:

Chatfield, C., 1989, The Analysis of Time Series: an Introduction. London, Chapman
and Hall, Fourth Edition, 241 p. (QA280.C4: a clear and concise treatment of all the
main topics.)

Diggle, P.J., 1990, Time Series: A Biostatistical Introduction. Oxford, Clarendon, 257
p. (QA280.D54 Clear introduction that includes spectral analysis as well as ARMA
models.)

Gottman, J.M., 1981. Time Series Analysis: a Comprehensive Introduction for Social
Scientists. Cambridge University Press, 400 p. (HA30.3 G67 : a well written
introduction to time domain analysis for readers without a strong math background.)

Pankratz, A., 1983, Forecasting with Univariate Box-Jenkins Models. New York, Wiley, 562 p. (QA280.P37 Simple introduction to time domain analysis with many examples of economic time series.)

More advanced treatment is given in:

Box, G.E.P. and G.N. Jenkins, 1994, Time Series Analysis: Forecasting and Control. Englewood Cliffs NJ, Prentice-Hall, third edition, 598 p. (QA280.B67 The standard work for ARMA models.)

Brockwell, P.J. and R.A. Davis, 1991, Time Series: Theory and Methods. New York, Springer-Verlag, second edition, 577 p. (QA280.B76)

Kay, Steven M., 1988, Modern Spectral Estimation: Theory and Application. Englewood Cliffs, NJ, Prentice-Hall, 543 p. (QA280.K39 Includes FORTRAN programs and a disk with data and source code.)

Marple, S.L., Jr., 1987, Digital Spectral Analysis with Applications. Englewood Cliffs, NJ, Prentice-Hall, 492 p. (QA280.H38)

Percival, Donald B., and Andrew T. Walden, 1993, Spectral Analysis for Physical Applications: Multitaper and Conventional Univariate Techniques. Cambridge University Press, 583 p. (QA320.P434 Available in paperback. Code in Lisp, and data sets are available by EMail.)

Press, William H. et al., 1989. Numerical Recipes in Pascal: the Art of Scientific Computing. Cambridge University Press, 759 p. (QA76.73.P2N87 Subroutines also available in Basic, FORTRAN, or C versions. See Chapter 12 for Fourier methods.)

Books representative of the Signal Processing literature include:

Bendat, J.S. and A.G. Piersol, 1971. Random Data: Analysis and Measurement Procedures. N.Y., Wiley-Interscience, 407 p. (TA340.B43 An excellent treatment of methods for time series with a large random component, such as turbulence measurements.)

Hayes, Manson H., 1996, Statistical Signal Processing and Modeling. New York, John Wiley and Sons, 608 p. (Includes MATLAB scripts.)

Newland, D.E., 1993. An Introduction to Random Vibrations, Spectral and Wavelet Analysis. Third Edition, John Wiley, 477 p. (Also lists program in FORTRAN for Fourier analysis, and MATLAB scripts for wavelets.)

Oppenheim, A.V., and A.S. Willsky, 1983, Signals and Systems. Englewood Cliffs NJ, Prentice-Hall, 796 p. (QA402.O63 A standard text.)

A readable book discussing atmospheric time series is;

Burroughs, W.J., 1992, Weather Cycles: Real or Imaginary? Cambridge University Press, 207 p. (QC883.2.C5B87)

The following apply time series analysis in the earth sciences:

Bennett, R.J., 1979, Spatial Time Series: Analysis—Forecasting—Control. London, Pion, 674 p. (QA280.B46 Many examples of both time and spatial series in the environmental sciences.)

Bras, R.L. and I. Rodríguez-Iturbe, 1985, Random Functions and Hydrology. New York, Dover, 559 p. (GB656.2.M33B73 An advanced text.)

Godin, G., 1972, The Analysis of Tides. University of Toronto Press, 264 p. (Theory of harmonic analysis, applied to tides.)

Haan, C.T., 1977, Statistical Methods in Hydrology. Ames IA, Iowa State University Press, 378 p. (GB656.2.S7H3 Chapter 14 gives a concise introduction to time series.)

Hegge, B.J. and G. Masselink, 1996, Spectral analysis of geomorphic time series: auto-spectrum. Earth Surface Processes and Landforms, v.21, p.1021–1040. (MATLAB functions are available by Email from `bruce@gis.uwa.edu.au`.)

Kinsman, Blair, 1965, Wind Waves: their Generation and Propagation on the Ocean Surface. Englewood Cliffs, NJ, Prentice-Hall Inc., 676 p. (GC211.K5 A classic work applying Fourier analysis to ocean waves, written in an informal style. Available in paperback.)

Munk, W.H. and D.E. Cartwright, 1966, Tidal Spectroscopy and Prediction. Phil. Trans. Roy. Soc. London, v. 299A, p.533-581 (A classic application of spectral analysis in oceanography.)

Robinson, E.A., 1981, Time Series Analysis and Applications. Goose Pond Press. (QA280.R64 Written by an expert in seismic signal processing.)

Schwarzacher, W., 1975, Sedimentation Models and Quantitative Stratigraphy. Amsterdam, Elsevier Publ. Co., 382 p.

Schwarzacher, W., 1987, Principles of quantitative lithostratigraphy. The treatment of single sections. Chapter IV.1 in Gradstein et al., Quantitative Stratigraphy (QE651.Q27 This article gives examples of correlograms and spectra calculated from stratigraphic sections.)

Chapter 8

Spatial Data

8.1 Introduction

As noted in the last Chapter, time series analysis may also be applied to one-dimensional spatial data, such as topographic profiles or stratigraphic sections. Most spatial data, however, are two or three dimensional. Only the more common, two-dimensional case is discussed in this book. The traditional way that earth scientists display such data is by using maps and contours. In the earth sciences, mapping and contouring software has reached a high degree of sophistication that cannot be matched by a general-purpose program such as MATLAB. The more advanced programs are now known as Geographical Information Systems (GIS). Nevertheless, MATLAB implements some routines that are useful for relatively small-scale applications.

The special problems of spatial data encountered by South African mining geologists in the 1950s, led in the 1960s and 1970s to the development of a theory designed for analysis of these data: it is now called *Geostatistics*. Unlike the term *Biostatistics*, which applies to any application of statistical methods to biological data, the term geostatistics applies to a particular methodology, given its present theoretical formulation by the French mathematician G.F. Matheron and his colleagues at the École Supérieure des Mines de Paris. Applications of geostatistics are no longer restricted to the mining industry: they have spread to the petroleum industry (Hohn, 1988; Yarus and Chambers, 1994), soil science, hydrology (Bras and Rodríguez-Iturbe, 1985), geohydrology, and many other fields. This chapter includes a brief introduction to these methods, and a few simple MATLAB routines. For large-scale applications, there are several packages available in the public domain (Deutsch and Journel, 1992, and see Yarus and Chambers, 1994, Chapter 25). For a broader approach to spatial statistics, see Cressie (1993).

132

8.2 Contouring

8.2.1 Introduction

A topographic contour is a line on a map connecting points of equal elevation. By extension, the term is use for lines on maps connecting points of equal value of any geographically distributed variable. Some authors prefer the term *isopleth* (or *isarithm*) for such lines, and a variety of other terms have arisen to describe particular applications. Examples include: *isobar* (equal atmospheric or other pressure); *isobath* (equal depth); and *isopach* (equal thickness of stratigraphic or other rock units).

Experience in the earth sciences has given rise to several rules, conventions, and styles of contouring. They have been nicely summarized in Tearpock and Bischke (1991, p.10–16):

There are five **rules** of contouring:

- Contour lines do not cross;

- Contour lines do not merge with other contour lines (though this, and crossing contours are theoretically possible where there is a vertical or overhanging cliff);

- Contour lines pass between control points with values higher and lower than the contour value;

- Contour lines are repeated to indicate a reversal in slope. Theoretically, it is possible that a slope might reverse exactly along a line equal to the contour value, but it is so unlikely that in drawing contours, it is never permitted;

- Contour lines must close, or end at the edge of the map.

The **conventions** of contouring are more flexible, but include the following: there should be a standard datum (e.g., sea level, for topographic contours); equality of contour intervals (which should be integral values, if possible); lines should be emphasized at a regular interval (e.g., every fifth line), and contours should be labelled for ease of reference.

Contouring by hand is a subjective process. This has both good and bad aspects. The person contouring can use his or her experience and background knowledge as a guide: for example, someone drawing a topographic map knows that streams are the lowest points locally, and that contours should 'V' upstream (a computer does not have this information), but a scientist's hopes or expectations may lead to drawing closure, or its absence, where it is not really justified by the available data. One way to minimize the subjective element is to follow a particular **style** of contouring. Two examples are:

- Parallel contours: here an attempt is made to keep the lines parallel, even though this increases their sinuosity. The spacing between lines may vary;

- Equally spaced contours: lines are both parallel and equally spaced. This corresponds to an extreme assumption about the data (all gradients are equal), and generally requires sharp bends in the lines.

The result of contouring by computer depends entirely on the particular algorithm used by the program. Generally, contouring takes place in three separate stages, each with its own set of algorithms.

- Gridding: irregularly spaced data must be converted to a set of points on a regular grid;

- Location of points: values at grid points are not the values required for contour lines, so points having these values must be found by interpolation;

- Line drawing: a technique must be chosen for connecting these points on a contour by a line. In advanced applications it is generally required that the line must be a smooth curve.

8.2.2   Contouring Using MATLAB

MATLAB provides a function `contour` to contour data on a regular grid, and saved as a matrix Z, and a function `mesh` to display such data as a three-dimensional block diagram. These are very useful tools for visualizing spatial data. A simple function `plotgrid` (not listed here) draws contours and labels such data (try it with `dgrid.dat`). Often data, particularly large data sets, are not saved in this form, but rather as a set of x, y, z values, where the x values are generally east-west coordinates (corresponding to the *columns* of a **Z** matrix), and the y values are north-south coordinates (corresponding to the *rows* of a **Z** matrix, but indexed from the bottom, not the top as MATLAB requires). The following function converts this type of file to an array, and displays it as (i) a contoured map; and (ii) a mesh diagram (note that this function only works correctly for data on a regular grid).

```
function [Z,t1,t2] = xyztogr(X, lab)
% [Z,t1,t2]  = xyztogr(X,lab)
% converts a matrix of x (E-W) coordinates, y (N-S)
% coordinates, and z values to a grid Z
% x and y need not be integers, or range from 1 up
% Z is a matrix of the gridded z values, t1 and t2
% are vectors which can be used as scales for the
% x and y axes. Also plots contour and mesh diagrams
% To label contours set lab to 1 (auto) or 2 (manual,
% using mouse), otherwise 0 (default)
% uses contvec to determine contour levels
% Written by Gerry Middleton, Feb 1997
if nargin < 2, lab = 0; end
eps = 0.000001;
x1min = min(X(:,1));  % determine maxima and minima for x (x1)
x1max = max(X(:,1));
x2min = min(X(:,2));  % and for y (x2)
x2max = max(X(:,2));
```

```
zmin = min(X(:,3));
zmax = max(X(:,3));
xx = sort(X(:,1));    % determine increments
inc1 = max(diff(xx));
xx = sort(X(:,2));
inc2 = max(diff(xx));
t1 = x1min:inc1:x1max+eps; % make x1 and x2 vectors
t2 = x2min:inc2:x2max+eps; % eps used to avoid rounding errors
ng1 = length(t1);     % find their size; ng1 is no of cols
ng2 = length(t2);     % ng2 is no of rows
ci = round((X(:,1) - x1min)/inc1 + 1); % convert to integer indices
ri = round((X(:,2) - x2min)/inc2 + 1);
Z = full(sparse(ri,ci,X(:,3))); % make array
v = contvec(zmin,zmax);
vc = length(v);
fprintf('Max value in data: %4.1f  ',zmax);
fprintf('Min value  %4.1f  \n',zmin);
fprintf('There are %2.0f equally spaced contour lines \n',vc);
fprintf('Press <Enter> to continue, at each pause \n');
fprintf('\n');
pause;
c = contour(t1,t2,Z,v);
if lab == 1
   clabel(c);
end
else if lab == 2
   clabel(c,'manual');
end
xlabel('x'), ylabel('y');
axis('equal');
pause;
figure
mesh(t1,t2,Z);
xlabel('X1'),ylabel('X2');
i = [ng2:-1:1];
Z = Z(i,:);   % reverse order of rows
```

The function first determines the two vectors t1 and t2 which contain the x and y scales. The lengths of these two vectors give the number of columns and rows of the grid. Next the x and y values are converted to integer indices, which are needed to use the MATLAB function (sparse) that converts the z values to a sparse matrix format: it is then converted into the full matrix format using full. See the Manual to see how these two functions operate. Next a suitable set of contour levels is determined by contvec.

```
function v = contvec(zmin,zmax)
% v = contvec(zmin,zmax)
% computes a suitable vector of contour
% intervals, given zmin and zmax
zdif = zmax - zmin;
lzd = floor(log10(zdif));
cdif = 10^(lzd-1);
s = 30;
cdif = cdif/2;
while s > 15
    cdif = 2*cdif;
    cmin = cdif*ceil(zmin/cdif);  % just above zmin
    cmax = cdif*floor(zmax/cdif); % just below zmax
    v = [cmin:cdif:cmax];
    s = length(v);
end;
```

This function selects between 7 and 15 rational contour levels, from a value just larger than the minimum value of z to a value just below the maximum level of z. Finally, xyztogr plots the contour map and the mesh diagram (see Fig. 8.1). Note that these may be printed *after* the function has run to the end, not while it is in pause mode. Also, if the default viewpoint used by mesh is not appropriate, it may be changed after the function has run to the end, by using the MATLAB function view. Figure 8.1 shows the results produced when this function is applied to the data in apxyz.dat, a set of data adapted from Isaacs and Srivastava (1989, Fig.2.1 on p.11). This same data is analyzed later in this Chapter, using trend analysis. Note that these figures were produced using MATLAB 4; the function also works in MATLAB 5, but produces slightly different figures. The appearance of the contours or grid can be improved in both versions of MATLAB by using interp2 to interpolate to a finer grid (see the MATLAB manual or help files).

MATLAB implements gridding of irregularly spaced points using the function griddata. The x and y coordinates are in vectors x and y, and the variable to be gridded z is in a vector z. The grid spacing is indicated by the input vectors xi and yi. xi must be a row vector (indicating the spacing of the columns), and yi a column vector (indicating the spacing of the rows). For earth scientists, the new version of griddata is probably the most significant improvement provided by MATLAB 5. The version used in MATLAB 4 interpolated the z values on the grid using an inverse distance method (based on all the data). For large data sets, this was slow and could exceed the computer's memory capacity. MATLAB 5 offers several different interpolation methods, with the default being linear interpolation from the "natural neighbors" (see discussion in the next section).

After the data have been gridded, they may be contoured. The number or values of contours plotted may be specified by a scalar n or a vector v respectively. The following

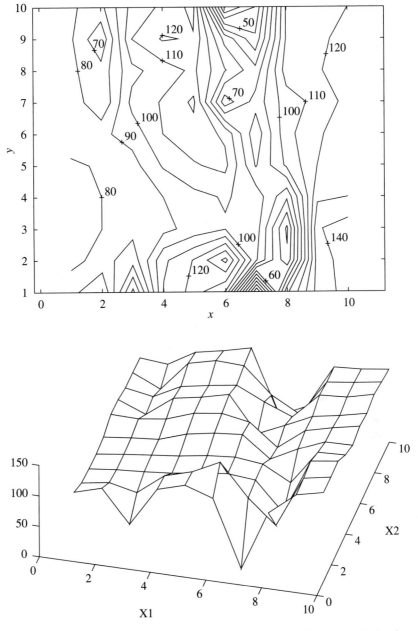

Figure 8.1   Contour map and mesh diagram of the topographic data set in Isaaks and Srivastava (1989, p.11)

MATLAB script draws a contour map of data previously loaded from the file `dmap.dat`. This file, taken from Davis (1986, p.362–363), contains 52 x, y, and z values as the three columns of 52×3 array.

```
ti = 0:0.25:6.5;
[xi,yi,zi] = griddata(dmap(:,1),dmap(:,2),dmap(:,3), ti, ti');
v = 700:25:950;
contour(ti, ti',zi, v), axis('square')
```

The result produced by MATLAB 4 is shown as the upper diagram in Figure 8.2. This data set has become a classic one, much worked over by other authors (e.g., Ripley, 1981). Davis (1986, p.377) provides the map, showing streams and hand-drawn contours, reproduced as the lower diagram in Figure 8.2. It is instructive to compare these maps, and the other examples of machine-drawn maps in Davis (1986) and Ripley (1981). As expected, the computed contours do not draw the contours correctly in the stream valleys, since the computer was not given the location of the streams, or the information that this is topographic data. MATLAB does not produce smoothed contours, and in fact uses the simplest of all line drawing algorithms (linear interpolation on a square grid, see following). The smoothness of the line can be improved by setting a finer spacing for the grid (e.g., 0.1 instead of 0.25). Another defect is the sharp drop off of the contours near the corners of the map. This effect is an "edge-effect" entirely due to the gridding algorithm used by MATLAB 4. It can be reduced or eliminated by gridding only *within* the range of the data set, i.e., by avoiding extrapolation, and the results produced by MATLAB 5 appear to do this.

MATLAB 4 also provided only a relatively crude method `clabel` for labelling contours, though it includes the (generally preferable) option of locating the labels using a mouse. MATLAB 5 adds an "inline" option to `clabel`, which further improves the appearance.

MATLAB provides a useful function `mesh` for examining surfaces in three dimensions. The following script produces a mesh diagram of the Davis topographic map surface, and plots the original data points on the same diagram;

```
mesh(xi,yi,zi);
view(170,30);
hold on;
plot3(dmap(:,1), dmap(:,2), dmap(:,3),'o');
hold off;
```

The result is shown in Figure 8.3. `view` sets the azimuth (170) and elevation angle (30) of view, in degrees. So the area is shown looking from just west of North, i.e., up the valley shown in Figure 8.2

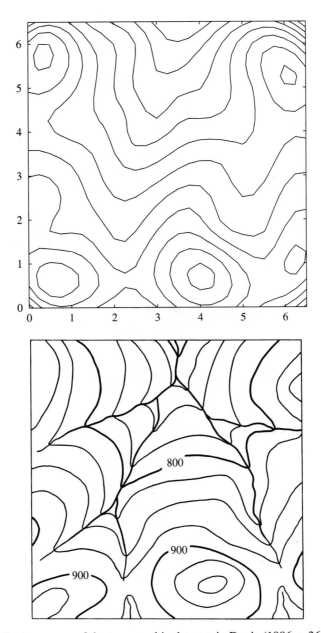

Figure 8.2   Contour maps of the topographic data set in Davis (1986, p.361–363): upper
map produced by MATLAB; lower map (from Davis, 1986, p.377) shows streams and
hand-drawn contours.

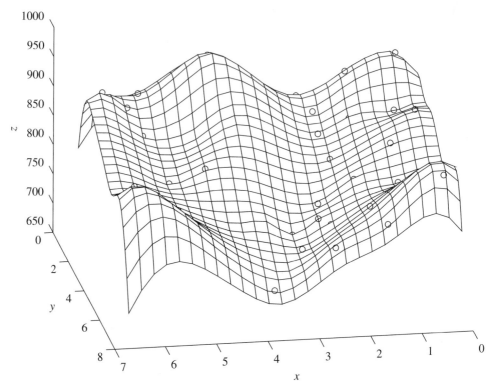

Figure 8.3   Mesh diagram of the surface contoured in Figure 8.2

8.2.3   Gridding and Contouring Algorithms

If the second stage in contouring is location of points on the contour by interpolation between known grid points, one might ask, why is gridding necessary? Why not simply interpolate between the original, irregularly spaced points (the control points)? Indeed it is possible to do this, but there are two arguments against it: (i) control points may be sparse in some areas, dense in others. Interpolation will therefore yield only a few points on the contour line in some areas, but many in others, which will present problems when the points are joined to make contour lines. (ii) Interpolation requires rational selection of the neighboring control points around every control point. It is not easy to give an algorithm that does this efficiently.

Let us consider briefly the second of these two problems. The problem of selecting the best neighbors for interpolation is a classic one, which has been solved independently in several different fields. It is the subject of a comprehensive monograph by Okabe et al. (1992). The "best" neighbors have been called the *natural neighbors*. Drawing lines between control points and their natural neighbors yields a set (*tesselation*) of triangles, known as *Delaunay triangles* that are unique, and completely cover the surface. This geometry also defines a polygonal area around (almost) every control point, such that every point in the polygon is closer to the control point than it is to any other control point. These polygons are called *Thiessen* or *Voronoi polygons*. Davis (1986, p.360) gives a simple geometric construction for selecting natural neighbors (see also Watson, 1992, p.59–63). Converting this geometric process into an efficient computer program has proved not to be an easy task. It has, however, been achieved and the algorithms are discussed at length in Okabe et al. (1992, Chapter 4).

An implementation of Delaunay triangles and Voronoi polygons is now available in MATLAB 5, and is the basis of the improved version of griddata. The following is a script to obtain the Delaunay triangles for the Davis map data.

```
load dmap.dat;
plot(dmap(:,1), dmap(:,2),'o');
axis([0 6.5 0 6.5],'square'), hold on
tri = delaunay(dmap(:,1), dmap(:,2));
    % tri lists the 3 indices of the data at each of the vertices
    % of all the triangles
trimesh(tri, dmap(:,1), dmap(:,2), dmap(:,3));
hold off, hidden off
```

For users of MATLAB 4, an implementation is included in a shareware package (SaGA) written by Kirill Pankratov available from the Web site:

```
http://puddle.mit.edu/~glenn/kirill
```

This package includes some 60 M-files for spatial analysis, and is a very valuable supplement to MATLAB 4 for those with a serious interest in this field. Among the program in SaGA is a simpler, faster triangulation program triangul and a simpler, faster version of griddata called griddat.

The triangulation of the Davis data produced using triangul is shown in the upper part of Figure 8.4. It is identical with the Delaunay triangulation produced by MATLAB 5, and essentially the same as that given by Davis (1986, p.364). The lower part of Figure 8.4 shows the result of contouring the output from Pankratov's griddat. The result is somewhat better than that produced by the MATLAB 4 griddata. The map also shows the result of adding contour labels manually, using MATLAB's clabel.

The simplest possible gridding algorithm is to determine the n control points closest to a grid point, and average their values of z. This can lead to bad results, particularly when all (or most) of the control points happen to lie on one side of the grid point. An unbalanced selection of neighbors can be avoided by a variety of stratagems, the best (but probably not the fastest) of which is to determine the natural neighbors of the grid points. Then it is generally better not to weight all neighboring control points equally, but to weight them as some inverse function of their distance, so that closer points are weighted more than those further away. MATLAB 5 offers a variety of weighting functions ("linear", "cubic", "nearest", and "invdist"). The first three options are based on interpolation from Delaunay triangles. The "nearest" method simply uses the value of the nearest vertex of the Delaunay triangle and is not generally suitable for contouring. The "cubic" method has the advantage

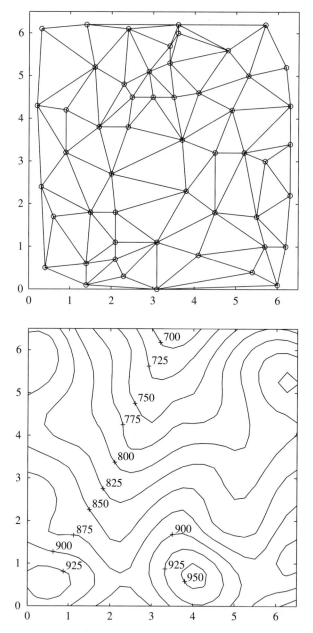

Figure 8.4   Upper map shows triangulation of Davis topographic data points produced using Kirill Pankratov's `triangul`. Lower map shows result of contouring output from Pankratov's `griddat`.

of producing a smooth interpolated surface (though it may not *look* smooth when contoured by MATLAB!).

An alternative, and perhaps more rational approach to weighting is *punctual Kriging* to be described in a later section of this Chapter.

The approach used by `griddata` in MATLAB 4 (or the "invdist" option in MATLAB 5) is more complicated. In Chapter 5 we described the use of cubic splines to fit a smooth curve to a one-dimensional set of control points. The technique can be extended to the two-dimensional case, where they are called *minimum curvature* or *biharmonic spline* methods. Note that, unlike methods based strictly on local weighted averages of nearby points, the surface fitted by spline techniques in theory depends on the complete set of control points, though in practice the local points exercise a much stronger control than the more distant ones. This feature of the method also means that the techniques are computationally intensive: large sets of equations must be solved to determine the weights a_i used to interpolate grid points. The interpolation equation has the form

$$z = \sum a_i g(r_i) \tag{8.1}$$

where the summation takes place over the total number of control points, and $g(r)$ is a function of the distance r_i between the interpolated point and the ith control point. This function is variously called a "basis function" or "Green's function." In MATLAB's application it is $g = r^2(\log(r) - 1)$, which is the theoretical value for elastic flexing of a thin sheet. As a result this version of `griddata` is slow, and may run out of memory space for large data sets. For further details of the technique see Watson (1992, p.122–123).

Pankratov's `griddat` subdivides the data set into subsets, and uses faster numerical routines. It also allows the user to choose the basis function to be used, although the function used by MATLAB remains the default. Some experimentation with other functions may (or may not) yield a better result. The function suggested by Pankratov $\exp(-r^2)$ actually gives much worse results with the Davis data set. The reader is encouraged to try out the various options for `griddata` available with MATLAB 5 by applying them to the Davis data set. This is facilitated by using the function `gridcon` included in the accompanying diskette.

Given a regular grid of values, the simplest algorithm for drawing contours is as follows:

- Set value of contour line, z_c;

- For each grid square, determine if the contour line intersects it (it must if $z_{min} < z_c \le z_{max}$, where z_{min} and z_{max} are the minimum and maximum of the values at the four corners of the square). If the contour does cut across the square, determine which sides it cuts, and where using a linear interpolation. Draw a straight line across the square from one intersection point to the other.

- Repeat for each square in the grid;

- Repeat for each contour value;

- End.

Some refinements are necessary to allow for the possibility that one of the grid points might have exactly the same z as the contour value, and so on. As we have seen, this simple method (used by MATLAB) does not produce a smooth contour, unless a very small grid spacing is used. Alternative linear interpolation methods use two or four triangles within each grid square: these methods produce somewhat smoother contours without requiring a finer grid. Alternatively, the contours may be smoothed by using some curve fitting routine, other than a straight line from one cut of the square (or triangle) to the other. This produces a more complex, slower program, which may produce crossing contours if special precautions are not taken to prevent this. Details of most of the methods that have been used are given in Watson (1992).

8.3   Trend Analysis

Trend analysis is the two dimensional equivalent of polynomial regression. A first order trend surface (called by some authors a "response surface") is a plane defined by the equation

$$y = b_0 + b_1 x_1 + b_2 x_2 \tag{8.2}$$

A second order trend surface is a curved surface, with one culmination or depression, defined by the equation

$$y = b_0 + b_1 x_1 + b_2 x_2 + b_3 x_1^2 + b_4 x_1 x_2 + b_5 x_2^2 \tag{8.3}$$

and so on. Such surfaces can be fitted to (x_1, x_2, y) data (also called (x, y, z) data) by multiple regression. The technique is first to generate the power and cross-product data from the original x_1 and x_2 data, and then use normal multiple regression techniques to generate the equation of the estimated trend surface. The technique is useful as a way of removing the "regional trend" from the data, in order to make it easier to see the "anomalies," i.e., the residuals $y_e = y - \hat{y}$. For example, for a stratigraphic surface such as an unconformity, the trend might consist of an imposed regional deformation (tilting and/or broad warping), and the residuals might reveal the original topography on the unconformity. In geophysical applications, there is often a regional trend in gravitational or magnetic fields due to deep seated structure, and local anomalies due to the properties of near-surface rock masses. The following function is a MATLAB implementation, for irregularly spaced data.

```
function b = trend(X,m,lab,ngrid)
% b = trend(X,m,lab,ngrid)
% Determines the trend surface of order m for a data matrix X
% with N rows and 3 columns: values of x1, x2, and y.
% Also plots contours on the original surface, the trend
% surface, and the residuals
% Set lab =1 for labels on contours
% ngrid is number of grid intervals, e.g., ngrid = 10
% generates 11x11 node grid, which is the default
% Uses polygen, cotomat, dattogr
% Written by Gerry Middleton, October 1996; modified June, 1998
```

```
if nargin < 4, ngrid = 10; end
if nargin < 3, lab = 0; end
[N,nc] = size(X);  % determine no of data (rows)
X1 = polygen(X(:,1:2), m);  % generate polynomial terms
X1 = [ones(N,1) X1 X(:,3)]; % expand by adding a first column of ones
S = X1'*X1;             % calculate sums of squares and products matrix
[n,n2] = size(S);
SS = S(1:n-1,1:n-1);   % normal equations are  SS*b = y
y = S(1:n-1,n);
b = SS\y;               % solution of normal equations
yhat = X1(:,1:n-1)*b;  % col. vector of expected values (y hat)
e = X1(:,n) - yhat;    % col. vector of residuals
x1min = ceil(min(X(:,1)));   % find limits of data
x1max = floor(max(X(:,1)));  % limits are just inside data
x2min = ceil(min(X(:,2)));   % to help curtail edge effects
x2max = floor(max(X(:,2)));
inc = 1/ngrid;                % generate grid coordinates
inc1 = (x1max - x1min)*inc;
inc2 = (x2max - x2min)*inc;
t1 = x1min:inc1:x1max;
t2 = x2min:inc2:x2max;
[Xi Yi] = meshgrid(t1,t2);
Z = griddata(X(:,1), X(:,2), X(:,3), Xi, Yi);
zmin = min(X(:,3));
zmax = max(X(:,3));
v = contvec(zmin,zmax);
vc = length(v);
fprintf('Max value in data: %4.1f  ',zmax);
fprintf('Min value  %4.1f  \n',zmin);
fprintf('There are %2.0f equally spaced contour lines \n',vc);
fprintf('Press <Enter> to continue, at each pause \n');
fprintf('\n');
pause;
c = contour(Z,v);
axis image; % change to axis('equal') for MATLAB 5
if lab == 1
   clabel(c);
end
xlabel('x'), ylabel('y');
title('Contours of Original Data');
pause;
yhmax = max(yhat);
yhmin = min(yhat);
figure;
```

```
fprintf('Max value in trend (at original pts): %4.1f  ',yhmax);
fprintf('Min value  %4.1f  \n',yhmin);
fprintf('\n');
T = cotomat(t1,t2);  % generate list of coords
Xt = polygen(T, m);  % generate polynomial terms
[rt ct] = size(Xt);
Xt = [ones(rt,1) Xt];
that = Xt*b;          % trend at each grid point
Y = dattogr([T that], ngrid+1, ngrid+1);
v = contvec(yhmin,yhmax);
c = contour(Y,v);
if lab == 1
   clabel(c);
end
axis image;  % change to axis('equal') for MATLAB 4
s = num2str(m);
title(['Contours on order ',s,' Trend']);
xlabel('x'), ylabel('y');
pause;
figure;
E = Z - Y;
emax = max(max(E));
emin = min(min(E));
v = contvec(emin,emax);
c = contour(E,v);
if lab == 1
   clabel(c);
end
axis image; % change to axis('equal') for MATLAB 4
title('Contours on grid residuals');
xlabel('x'), ylabel('y');
pause;
fprintf('Max value in residuals (at original pts): %4.1f  ', max(e));
fprintf('Min value  %4.1f  \n',min(e));
fprintf('\n');
```

Note that this function uses contvec, described in the previous section, and three other utilities:polygen, which generates the power and cross-product terms, cotomat, which generates a $(mn \times 2)$ matrix from a set of grid coordinates (m x values, and n y values), and dattogr which converts a set of (x, y, z) values to a grid of z values. The M-files are included in the diskette that accompanies this book. The diskette also includes a more elaborate version trend3 that will accept and plot data that are already regularly spaced. Figure 8.5 shows the result of applying this version to the data from Isaaks and Srivastava (1989, p.11), displayed in the previous section.

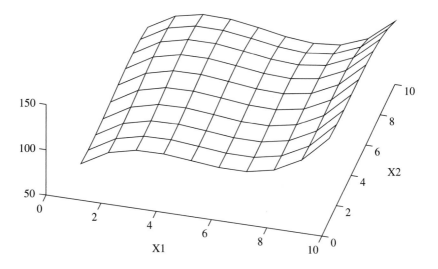

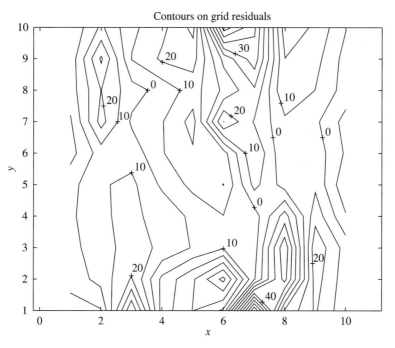

Figure 8.5   The upper diagram is a mesh diagram of the third order trend surface on the topographic data from Isaaks and Srivastava (1989, p.11), and the lower diagram shows contours on the residuals from this surface. Compare with Figure 8.1.

The significance of the trend surface fit can be assessed from the analysis of variance table printed out by `mreg`. Note that the ordering of the variables (`x1`, `x2`, `x3`, `x4`, `x5`, . . .) is the same as that shown in Equation 8.3. The results (not given here) show that the third order trend is statistically significant, even though it cannot adequately represent the complex topography of this area near the California–Nevada border, as revealed by the large residuals.

8.4 Geostatistics

8.4.1 The Semivariogram

By now, we are familiar with the notion that spatial data tend to be autocorrelated, with the highest correlation between closely spaced stations. In time series analysis, the way that the correlation varies with spacing is measured by the autocorrelation or autocovariance. In geostatistics it is measured by a function that is closely (but inversely) related to the autocovariance: the *semivariogram*. The semivariogram γ_h for a spatial series x_i, $i = 1, 2, \ldots N$ with a regular spacing between the x values of Δx, is given by

$$\gamma_h = \frac{1}{2N} \sum_{i=1}^{N-h} (x_i - x_{i+h})^2 \tag{8.4}$$

The relation between the semivariance and the autocovariance c_h is straightforward (Carr, 1995, p.161–162)

$$\gamma_h = c_0 - c_h \tag{8.5}$$

where c_0 is the variance. The practice of using the semivariance in geostatistics, rather than the autocovariance, is just the result of the historical development of the discipline. Some authors multiply the semivariogram by two, and call it the *variogram*.

The semivariogram shows the way that the sum of squares of the differences between stations (scaled by $2N$) varies with the distance between the stations $h\Delta x$. We expect that there will be a minimum C_o at (almost) zero spacing, and that the semivariance will level out at some maximum value $C_o + C$, called the *sill*, for h values larger than a certain value, called the *range*. Between zero spacing and the range, there are several different models for how the semivariance changes: a few are shown in Figure 8.6.

Notice one important difference between these models and an autocovariance model: the semivariogram allows for the possibility that, at near-zero spacing, there is still some difference between sample values. Of course, if we simply calculate the semivariance using Equation 8.4 the zero spacing value must be zero; but if we consider that the value at a spacing of near-zero should correspond to the result we would obtain by resampling "the same" station, then we realize that it is generally not zero because of local variation and the practical impossibility of ever taking exactly the same sample again. This local variation is called the "nugget effect" in geostatistics, because it is an effect seen most prominently in assays made for a substance like gold, which tends to occur in small, widely separated, nuggets. Assays of samples taken very close together (i.e., at near-zero spacing) therefore

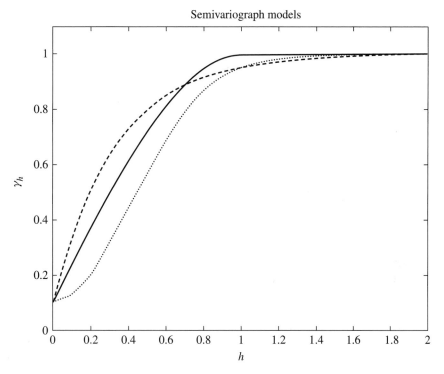

Figure 8.6 Model semivariograms. The models are: full line – spherical; dashed line – exponential; dotted line – Gaussian.

tend to have a variance substantially different from zero. The nugget value is not derived by direct calculation from the measured values x_i, but from the model that is fitted to the measured values.

The equations for these models are:

- Spherical model

$$\gamma_h = \begin{cases} 0 & \text{if } h = 0 \\ C_o + C(1.5(h/a) - 0.5(h/a)^3) & \text{if } 0 < h \leq a \\ C_o + C & \text{otherwise} \end{cases} \quad (8.6)$$

Note that the spherical model looks very much like two straight lines, with a smooth curve connecting them: at small values of h/a the first (linear) term is dominant, and the second (cubic) term becomes important only for values of h/a almost equal to one.

- Exponential model

$$\gamma_h = \begin{cases} C_o + C(1 - \exp(-|h|/a)) & \text{if } h > 0 \\ 0 & \text{otherwise} \end{cases} \quad (8.7)$$

- Gaussian model

$$\gamma_h = \begin{cases} C_o + C(1 - \exp(-(h/a)^2)) & \text{if } h > 0 \\ 0 & \text{otherwise} \end{cases} \tag{8.8}$$

Notice that the spherical model is essentially a polynomial in h/a, and the exponential and Gaussian models can be reduced to a linear equation by a logarithmic transformation. This means that model curves can be fitted to real data using the regression techniques discussed in Chapter 5. In practice, one must first estimate a. There is also some question about the value of a to use in the equations for the Exponential and Gaussian models, as the sill is approached asymptotically. In plotting Figure 8.6, we have used the suggestion by Journel and Huijbregts (1978, p.164–165) that a practical value of a can be set at the value a' where γ reaches 95% of the ultimate sill value. This is achieved by replacing a in the equations by $a'/3$ and $a'/\sqrt{3}$ respectively.

Before giving an example of a theoretical fit to data, however, we have to discuss how the semivariogram is calculated from real data, which generally are irregularly spaced map data rather than regularly spaced data measured on a linear traverse.

The problem of estimating the semivariogram from real data is discussed in many texts on geostatistics (e.g., Isaaks and Srivastava, 1989; Carr, 1995, Chapter 6; Deutsch and Journel, 1992). Carr (1995, p.167) presents an algorithm. The following is a brief summary. It is assumed that there is no regional trend—if there is, it should first be removed by using trend analysis. First choose a class size h, and set a limit to the number of increments to be evaluated m. Then for each of the points on the map (stations) for which there is data calculate distances to every other station, and classify each pair in one of the distance classes ($h\Delta x$, $2h\Delta x$, ..., $mh\Delta x$) and also in a directional class (e.g., azimuth classes 0–19,20–39,...,160–179). This involves an approximation: the distances and azimuths are classified into classes, so they have approximately that spacing, along approximately that spatial direction, rather than having an exact spacing along a particular traverse. Once the data are all classified, a *directional semivariogram* may be calculated for each direction. If these semivariograms show no strong differences (i.e., the semivariance is isotropic) the data may be averaged to give a single plot.

From this description, it is apparent that the number of numerical operations is quite large. Pankratov has given an efficient MATLAB implementation as the function `kriging` in his SaGA toolbox. Instead of duplicating his efforts, we present the following much simpler implementation for data on a regular grid.

```
function gam = semivxy(X, h)
% gam = semivxy(X,h)
% Function to calculate the semivariogram of a
% regular array X measured in the x and y direction
% (along the rows and columns of the array)
% it is assumed that the x and y spacing are equal
% h is the number of lags over which it is calculated
% Written by Gerry Middleton, November 1995.
[r c] = size(X); %r is no of rows, c no of cols
```

```
for i = 1:r %for each row in turn
   for j = 1:h
      xx = X(i,1:c-j); %data from 1st to (c-j)th col
      y = X(i,1+j:c);  %data lagged by j to end of col
      G(i,j) = sum((xx-y) .^ 2)/(2*(c-j));
      end
   end;
gamx = sum(G)/r;
for i = 1:c %for each col in turn
   for j = 1:h
      xx = X(1:r-j,i); %data from 1st to (r-j)th row
      y = X(1+j:r,i);  %data lagged by j to end of row
      G(i,j) = sum((xx-y) .^ 2)/(2*(r-j));
                       %each row has gamma for a col
   end
end;
gamy = sum(G)/c;  %sum the cols and average
plot(0:h,[0 gamx],'o', 0:h, [0 gamx], ':',...
   0:h, [0,gamy],'+',0:h, [0 gamy],':');
gam = (gamx + gamy)/2;
```

The upper part of Figure 8.7 shows semivariograms calculated for the 24×24 data grid in file brooker.dat, taken from Brooker (1991, Appendix A, p.72–73). The lower part shows a spherical model fitted to the combined x and y data using the following function.

```
function [b] = svfit(gam, a)
% [b] = svfit(gam, a)
% svfit fits a spherical semivariogram model to the
% empirical semivariogram data in the row vector gam,
% using polynomial regression on the first a values.
% Written by Gerry Middleton, December 1996
n = length(gam);
if n < a
   error('a too large');
end
x = [1:a];
x = (x ./ a);
X(:,1) = x';
x3 = (x .^3);
X(:,2) = x3';
X(:,3) = gam(1:a)';
m = mean(X);
D = X - ones(a,1)*m; % convert X to deviations from mean
SS = D'*D;
                    % partition the matrix SS
```

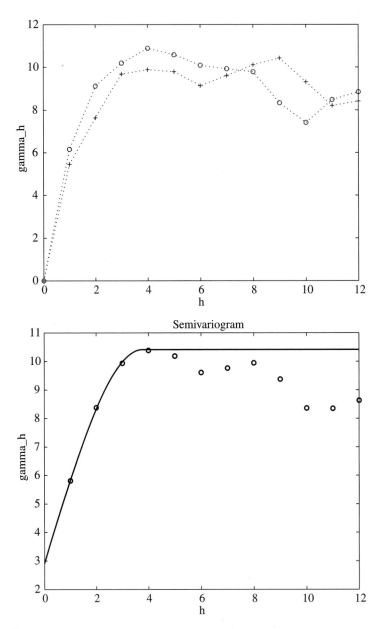

Figure 8.7   Upper diagram shows semivariograms calculated for a regular data grid;
circles are for the x direction, and crosses for the y direction. Lower diagram shows
spherical model fitted to combined x and y data.

```
S = SS(1:2, 1:2);
P = SS(1:2,3);
t = SS(3,3);
             % solve for regression on the first variable
b1 = P(1)/S(1,1);
r  = P(1)/sqrt(t*S(1,1));
SR1 = r*r*t;
                % now do the second
SI = S(1:2,1:2);    %redefine S,B,SR
PI = P(1:2);
BI = SI\PI;
SRI = BI'*PI;       %sum of squares due to this regression
SRA = SRI - SR1;    %sum of squares added by this regression
fprintf('SS due to regression on x1 = %8.2f\n',SR1);
fprintf('SS added by x2              = %8.2f\n',SRA);
fprintf('Total SS                    = %8.2f\n\n',t);
b0 = m(3) - BI(1)*m(1) - BI(2)*m(2);
b = [b0; BI];
x1 = (0:0.1:a) ./ a;
x3 = x1 .^3;
f = b(1) + b(2)*x1 + b(3)*x3;
fmax = max(f);
k = find(f == fmax);
x = 1:n;
plot(x, gam,'o',a*x1(1:k),f(1:k), [a*x1(k) n],...
    [f(k) f(k)],'LineWidth',1);
title('Semivariogram');
xlabel('h'), ylabel('gamma_h');
fprintf('Nugget Co = %8.2f\n',b(1));
fprintf('Sill C+Co = %8.2f\n',fmax);
```

To use this function one must choose a value of h, which can be regarded as a tentative estimate of the range a. Then the function determines the regression of γ_h, on (h/a) and $(h/a)^3$ using a modification of the mreg function presented in Chapter 5. From this regression, one can calculate the maximum value of γ_h, which is the sill value, and the intercept on the γ_h axis, which is the nugget value. Some experimentation, using different input values of a, may be necessary before a good fit is obtained.

The function presented uses ordinary least square techniques to fit the model to the data. It is better, however, to use a nonlinear regression technique that weights the data inversely by spacing, so that the fit at low spacings is determined more accurately than the fit at high spacings. See McBratney and Webster (1986), Cressie (1993), or Kitanidis (1997, Chapter 4) for details of these, and other, complications in the techniques of estimating semivariograms. One good technique that might be used is the simplex routine discussed in Chapter 5. Almost all authors agree, however, that computed semivariograms are not very good estimates of population semivariograms, and that fitting model semivariograms is somewhat subjective, no matter what technique is used. For this reason, many prefer to fit the semivariogram model by simple trial-and-error.

8.4.2 Kriging

What use is the semivariogram, once it has been determined? The answer is that it is used for interpolating unknown values from measured values, a technique known as *kriging* after the South African mining engineer, D.G. Krige, who first developed it as a practical method (for a brief history, see Carr, 1995, p.151–152). Instead of doing this by using arbitrary polynomials or splines, it can now be done using a function (the semivariogram) "known" to be valid in the area of interest. The quotation marks remind us that, in reality, the form of the semivariogram is generally not very well known, with the data showing large deviations from the theoretical models that are used. Nevertheless, the practical value of using the geostatistical approach is attested by its popularity in the mining industry, where large excavation costs can be reduced substantially by correctly predicting the grade of ore that has not yet been sampled or mined.

Two different types of prediction are distinguished.

- **Punctual Kriging** predicts the value of the spatial variable at particular points. It is therefore a useful technique for gridding irregularly spaced data, so that they may be contoured.

- **Block Kriging** predicts the average value of the spatial variable over a specified area. It is a useful technique in mining, where decisions about whether or not to excavate a given "block" of ore depend upon estimating its grade.

In this book, we confine our attention to punctual kriging: for block kriging see Carr (1995, p.188–204, and references listed therein).

In punctual kriging, we first select the closest measured neighbors to the unmeasured point of interest. The number of neighbors selected is still somewhat arbitrary, but not entirely so, because we now know the range, which estimates the distance around the unmeasured point where the most useful data for prediction are to be found. If there are no neighbors closer than the range, then we might as well estimate the unmeasured point by using the regional average. If there are neighbors within the range, then using a weighted average of those data should yield a better estimate. Kriging, therefore, is a technique that uses the semivariogram to determine the best weights w_i to use in the equation to predict z_p from a set of known values z_i

$$\hat{z}_p = \sum_{i=1}^{n} w_i z_i \tag{8.9}$$

We expect that the prediction will not be completely accurate but will differ from the true value by a random error ϵ_p. An important advantage of kriging is that it yields not only the estimated value, but an estimate of the error variance s_ϵ^2.

The prediction itself is based on a least squares procedure, which yields the following set of equations

$$w_i \gamma(h_{ij}) = \gamma(h_{ip}) \tag{8.10}$$

where w_i is the unknown vector of weights, $\gamma(h_{ij})$ is a square matrix of semivariances calculated from the distances h_{ij} between all pairs of the known control points, and $\gamma(h_{ip})$ is a vector of semivariances calculated from the distances h_{ip} between the control points and the unmeasured point. In addition, there is the constraint that the weights must sum to unity. The set of equations is therefore overdetermined, and must be solved using a *Lagrangian multiplier* λ. In matrix form, we can represent the equations to be solved as

$$\begin{bmatrix} \gamma & k \\ k' & 0 \end{bmatrix} \begin{bmatrix} w \\ \lambda \end{bmatrix} = \begin{bmatrix} \gamma_p \\ 1 \end{bmatrix} \tag{8.11}$$

where k is a column vector of ones. The notation is the same as in the previous equation except that we have omitted the subscripts i and j. The error variance turns out to be estimated by

$$s_\epsilon^2 = \sum w_i \gamma(h_{ip}) + \lambda. \tag{8.12}$$

Useful worked examples are given by Davis (1986, p.386–393), Isaaks and Srivastava (1989, p.290–313), and Carr (1995, p.184–187). In practice, the covariance matrix is often used instead of the semivariance matrix. The following is a MATLAB function that performs punctual kriging. The input data consists of a matrix M giving x, y, and z for the control points, a vector v giving x and y for the point whose value z is to be estimated, and the coefficients C_o, C, and a for the semivariogram model that is to be used.

```
function [z, s2] = pkrige(M, v, co, c, a, mod)
% [z, s2] = pkrige(M, v, co, c, a, mod)
% returns a punctual kriging estimate of the value z
% and the error variance s2 of an unknown point whose
% xy coordinates are given by the row vector v
% M is the n by 3 matrix of x,y,z values of the
% n control points. co, c and a are the parameters
% of the fitted semivariogram, and mod is the
% model used: mod = 1: spherical; mod = 2: exponential
% mod = 3: Gaussian.
% Written by Gerry Middleton, December 1996
[n col] = size(M);
X = [M(:,1:2); v];
H = zeros(n+1); % calculated distance matrix (see Edist)
for i = 1:2
   X1 = X(:,i)*ones(1,n+1);
   DX = X1-X1';
   H = H + (DX .^2);
end;
H = sqrt(H);
```

```
Gam = zeros(n+1);
if mod == 1              % calculate semivariance matrix
   for i = 1:n+1
      for j = 1:n+1
         if i == j
            Gam(i,j) = 0;
         elseif H(i,j) > a
            Gam(i,j) = co + c;
         else
            h = H(i,j)/a;
            Gam(i,j) =  c*1.5*h - 0.5*h^3;
         end
      end
   end
   fprintf('Semivariances using spherical model\n');
elseif mod == 2
   Gam = co + c*(1 - exp(-3*abs(H/a)));
   fprintf('Semivariances using exponential model\n');
elseif mod == 3
   Gam = co + c*(1 - exp(-(sqrt(3)*(H/a)) .^2));
   fprintf('Semivariances using Gaussian model\n');
else
   error('Wrong entry for mod: should be 1,2, or 3');
end;
Gam     % comment out this line if no printout required
gamp = Gam(:,n+1);     % vector gamma(p)
gamp(n+1) = 1.0;
Gam(:,n+1) = [ones(n,1); 0];  % modify Gam
Gam(n+1,1:n) = ones(1,n);
w = Gam\gamp;           % weights, last term is lambda
z = w(1:n)'*M(:,3);    % predicted value
s2 = w'*gamp;
plot(M(:,1), M(:,2),'o');  % comment the following lines
hold on;                    % out if no plot is required
plot(v(1), v(2), '+');
axis('equal');
dx = 0.02*(max(M(:,1)) - min(M(:,1)));
for i = 1:n
   text(M(i,1)+dx, M(i,2), num2str(M(i,3)));
end
text(v(1)+dx, v(2), num2str(z));
hold off
```

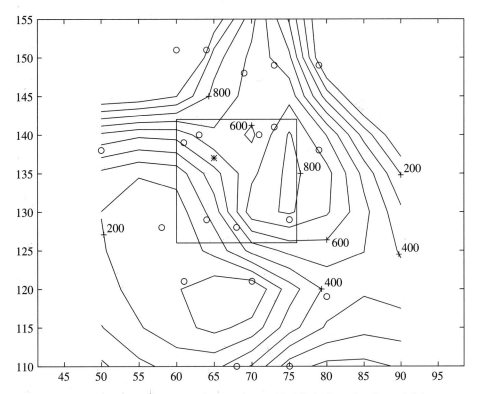

Figure 8.8   Example of interpolation by punctual kriging, from Isaaks and Srivastava
(1989, p.291–292)

The result of applying this function to an example data set, taken from Isaaks and
Srivastava (1989, p.291–292) is shown in Figure 8.8. The estimated point (with coordinates
[65, 137]) is shown by the asterisk, and the data used to estimate it came from the seven
points enclosed in the small box (see the file iskrig.dat). The contour map was draw using
MATLAB's routines, applied to the data points plotted (taken from Table 6.1 in Isaaks and
Srivastava, 1989, p.115–119; see the file iskrig2.dat). The exponential semivariogram
model was used, with $C_0 = 0$, $C = 10$, $a = 10$. The numerical results are identical with
those given by Isaaks and Srivastava: the estimated z is 592.7, and the error variance s_ϵ^2 is
8.96. If we had used only the contours to estimate this point, the estimate would probably
have been somewhere in the range 500–600, and it would have been difficult to have been
precise about its possible error. The reader can experiment with the effect that a different
choice of model has on this result.

8.4.3   Conclusion

In this Chapter we have provided only a few example applications of MATLAB to two
large, complex, partly inter-related topics: contouring and geostatistics. There is much

more to both of these subjects than we have been able to discuss: see, for example, the books by Watson (1992) and Deutsch and Journel (1992). One extension of kriging, called *cokriging*, is the use of other spatial measurements to help predict the variable of interest. For example, in a study of a particular oil reservoir or water aquifer, porosity may have been measured in only a few wells, but borehole geophysical properties were measured in many more. Cokriging attempts to use not only the porosity measurements, but also the geophysical variables to predict porosity at wells where it was not measured. An extensive MATLAB implementation of cokriging is given by Marcotte (1991). Marcotte (1996) also shows how MATLAB may be used for fast computation of variograms from large, regularly spaced data sets, such as those resulting from remote sensing.

Geostatistics is based on an extensive theoretical foundation, which we have not attempted to present. Some of its proponents have made claims for it which other theoreticians and practitioners are not able to accept. For the flavor of these sometimes acrimonious disputes see Philip and Watson (1986) and Journel (1986). The following quotation from Deutsch and Journel (1992, p.4) is representative of the more modest claims made for geostatistics by some of its proponents:

> Most early applications of geostatistics were related to mapping the spatial distribution of one or more attributes, with emphasis given to characterizing the variogram model and using the kriging (error) variance as a measure of estimation accuracy. Kriging, used for mapping, is not significantly better than other deterministic interpolation techniques that are customized to account for anisotropy and other important spatial features of the variable being mapped...Unfortunately, the kriging variance is independent of the data values and cannot be used, in general, as a measure of estimation accuracy.

In this book, no attempt has been made to discuss the more recent applications of geostatistics to modeling—for examples in petroleum and groundwater geology see Yarus and Chambers (1994).

There is also much to spatial statistics, besides contouring and geostatistics (e.g., properties of point patterns, whether they are random or clustered, etc.). For a complete treatment, see Cressie (1993). We consider directional properties of spatial data in the next Chapter.

8.5   Recommended Reading

Brooker, Peter I., 1991, A Geostatistical Primer. Teaneck NJ, World Scientific, 95 p. (QE33.2.S82B75 Brief introduction, with minimal theory.)

Carr, J.R., 1995, Numerical Analysis for the Geological Sciences. Englewood Cliffs NJ, Prentice-Hall, 592 p. (QE33.2.M3C37 Chapter 6 gives a clear introduction to geostatistics, and Chapter 7 discusses contouring and wireline diagrams: with FORTRAN programs.)

Cressie, N.A.C., 1993, Statistics for Spatial Data. N.Y., John Wiley and Sons, revised edition, 900 p. (QA278.2.C75 The most comprehensive text, but does not discuss contouring.)

Harbaugh, J.W. and D.F. Merriam, 1968, Computer Applications in Stratigraphic Analysis. New York, John Wiley and Sons, 282 p. (QE652.H3 Now rather out-of-date, but gives clear discussion of some contouring algorithms.)

Hohn, M.E., 1988, Geostatistics and Petroleum Geology. New York, Van Nostrand Reinhold, 264 p. (TN871.H536 A clear introduction, with many examples.)

Isaaks, E.H. and R.M. Srivastava, 1989, An Introduction to Applied Geostatistics. Oxford University Press, 561 p. (QE33.2.M3I83 Uses minimal mathematics and a case-history approach.)

Jones, T.A. and D.E. Hamilton, 1986, Contouring Geologic Surfaces with the Computer. New York, Van Nostrand Reinhold, 314 p. (QE36.J66 Discusses general problems, but gives no algorithms.)

Journel, A.G., 1986, Geostatistics: Models and tools for the earth sciences. Mathematical Geology, v.18, p.119–140 (in part a rebuttal of the paper by Philip and Watson, 1986.)

Kitanidis, P.K., 1997, Introduction to Geostatistics: Applications in Hydrogeology. Cambridge University Press, 249 p. (GB1001.72.S7K57 Includes a few MATLAB programs.)

Lancaster, P. and K. Šalkauskas, 1986, Curve and Surface Fitting: An Introduction. New York, Academic Press, 280 p. (QA297.6 Discusses surface fitting with polynomials, blending methods, finite elements, moving least squares, and splines.)

Marcotte, D., 1991, Cokriging with MATLAB. Computers and Geosciences, v.17, p.1265–1280 (with an extensive set of MATLAB functions).

Marcotte, D., 1996, Fast variogram computation with FFT. Computers and Geosciences, v.22, p.1175–1186 (uses MATLAB).

McBratney, A.B. and R. Webster, 1986, Choosing functions for semi-variograms of soil properties and fitting them to sampling estimates. Journal of Soil Science, v.37, p.617–639.

Philip, G.M. and D.F. Watson, 1986, Matheronian geostatistics—Quo vadis? Mathematical Geology, v.18, p.93–117. (An attack on geostatistical methods: see Journel, 1986, for a rebuttal.)

Ripley, B.D., 1981, Spatial Statistics. New York, John Wiley and Sons, 252 p. (QA278.2.R56)

Tearpock, D.J. and R.E. Bischke, 1991, Applied Subsurface Geological Mapping. Englewood Cliffs NJ, Prentice-Hall, 648 p. (TN870.5.T38 Emphasis on older, non-computer methods.)

Watson, D.F., 1992, Contouring: A Guide to the Analysis and Display of Spatial Data. Oxford, Pergamon Press, 321 p. (GA125.W38 With a large bibliography, and a set of BASIC programs. An essential reference for anyone seriously interested in contouring.)

Yarus, J.M. and R.L. Chambers, eds., 1994, Stochastic Modeling and Geostatistics: Principles, Methods, and Case Studies. American Association of Petroleum Geologists Computer Applications in Geology, No. 3, 379 p. (A collection of papers explaining how geostatistics can be applied in the petroleum and groundwater fields: emphasis on its use for modeling reservoirs.)

Chapter 9

Directional and Compositional Data

9.1 Introduction

This Chapter is devoted to two rather different types of data, both very important in the earth sciences, and rarely treated in introductory statistics texts or courses.

- Directional data are data that measure orientation in two or three dimensions. Two dimensional data, also called *circular* data are orientations in a plane, for example, the horizontal components of current directions, and their magnitudes, or the geological record of such currents, called *paleocurrents*. Paleocurrents are recorded by sedimentary structures, and are generally measured in the plane of bedding, which was originally approximately horizontal. They include structures, such as cross-bedding and flute casts, that indicate the *direction of movement* (so are periodic on a 360 degree scale), and also structures or textures, such as lineations and grain orientations measured in the plane of bedding, that only indicate the *line of movement* (so are periodic on a 180 degree scale). These two types of data are called *vectorial* and *axial* by Fisher (1993).

  Other data, not so obviously "circular" but sharing much of the regular periodicity of directional data, are measurements related to a particular time of day or other time cycle, for example, the daily variation in temperature, or the yearly variation in atmospheric carbon dioxide.

  Three dimensional data include orientation of currents measured in three dimensions, orientation of geological structures (such as bedding, schistosity, or metamorphic lineations) and textures (such as morphological orientations or crystallographic orientation of grains): they too may be either directional or linear in character.

- Compositional data are data measured on a percentage (or parts per thousand, or parts per million) scale. They are measured on a scale that varies only from zero to 100 percent, and they too present special problems for graphical display and statistical analysis. In fact, there are few accepted techniques for statistical analysis of these *closed array* data.

Why consider both types of data in a single Chapter? Compositional data are often plotted on triangular diagrams, where the three variables plotted (say X, Y, and Z), are constrained to sum to 100 percent, i.e., recalculated to

$$x \;=\; X/(X+Y+Z) \tag{9.1}$$
$$y \;=\; Y/(X+Y+Z) \tag{9.2}$$
$$z \;=\; Z/(X+Y+Z) \tag{9.3}$$

Philip and Watson (1988b) have pointed out that such data could equally well be recalculated to a vector of unit length, extending from the origin to x', y', and z', where

$$x' \;=\; \sqrt{X^2/(X^2+Y^2+Z^2)} \tag{9.4}$$
$$y' \;=\; \sqrt{Y^2/(X^2+Y^2+Z^2)} \tag{9.5}$$
$$z' \;=\; \sqrt{Z^2/(X^2+Y^2+Z^2)} \tag{9.6}$$

In this case, they are essentially directional data, restricted (for positive compositions) to a single 90 degree quadrant of three-dimensional space.

9.2   Graphical Displays of Directional Data

In the earth sciences, the most popular way to display directional data is as a *rose diagram*, if the data are circular, or on a *stereographic projection*, if the data are spherical. Neither of these techniques is well implemented by existing functions in MATLAB. MATLAB has a rose function, but it is not useful for displaying circular data in the form preferred by earth scientists.

Before functions are described that *will* display directional data in the forms preferred by earth scientists, we must take note of the different conventions for angles adopted by most scientists (and by MATLAB), and those almost universally adopted by earth scientists.

Earth scientists generally measure circular orientation by means of an azimuth, in degrees, measured clockwise from North. Other scientists generally measure angles, in radians, anticlockwise from the positive x-axis direction. On a map, this would be East. To convert from degrees to radians use

```
rad = pi*deg/180
```

and to convert from the earth science convention to the MATLAB convention use

```
theta = pi*(90 - az)/180
```

So, for example, an azimuth of 120 degrees is converted to a theta of -30 degrees, or -0.524 or $-\pi/6$ radians. All angles used in MATLAB must be expressed in radians.

In three dimensions, earth scientists generally express orientation using a vector consisting of a dip and an azimuth. Although geologists are sometimes taught to use the dip angle and the strike direction, this is an outdated, potentially confusing convention. It is better to use the angle of dip, and azimuth of the dip direction (both measured in degrees). In the earth sciences, dip angles are positive if down, negative if up (supposing that the distinction can be made). If we wish to express such a vector (of length L) as an array (x, y, z) in the usual, righthanded coordinate system (with the xy-plane horizontal, y North, and z vertically up), then the transformation is

$$x = \text{L} \cos \text{dip} \sin \text{az} \qquad (9.7)$$
$$y = \text{L} \cos \text{dip} \cos \text{az} \qquad (9.8)$$
$$z = -\text{L} \sin \text{dip} \qquad (9.9)$$

Alternatively, the usual spherical angles θ and ϕ may be calculated (in degrees) from

$$\theta = \text{dip} + 90 \qquad (9.10)$$
$$\phi = 360 - \text{az} \qquad (9.11)$$

(Derivations are given by several texts, including Middleton and Wilcock, 1994, p.60.)

9.2.1 Rose Diagrams

Rose diagrams are a form of circular bar chart or histogram (Fisher, 1993, points out that they were first used by Florence Nightingale in 1858). To be a true histogram, the frequency must be represented by an area. The area of a rose diagram segment is proportional to the square of the radial scale, so for a rose to be a histogram the radial scale must be proportional to the square root of the radius. This has the further advantage that it avoids visual exaggeration of the modal classes. Figure 9.1 shows examples of the two types of diagram. The data are orientations of feldspar laths in a lava flow (in file `ffeld.dat`, taken from Fisher, 1993, Appendix B2 on p.240), so they are an example of a "line of movement" data set. The figures were produced by the following MATLAB functions:

```
function ff = grose3(az, nb, typ, fscale)
% ff = grose3(az, nb, typ, fscale)
% This is a replacement of MATLAB's rose
% for those in the geophysical sciences.
% It draws a rose diagram of the azimuths (in degrees)
% in the vector az, after classifying them into
% nb classes. typ is set to 1 (the default value)
% for 360 degree data, or 2 for 180 degree data.
% set fscale = 0 for arithmetic, 1 for square root
% scale for frequency.
% This version uses a modified version of centaxes
% (centax2) to plot a central scale: to turn this
% off, comment out the call to centax2.  The
```

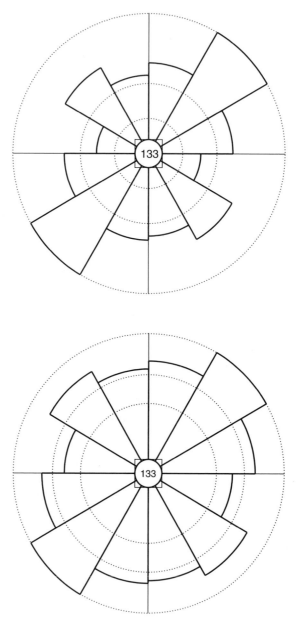

Figure 9.1   Two rose diagrams of the same data: the upper diagram uses a linear frequency scale, the lower diagram uses a square root scale. The dotted circles indicate the maximum frequency, and frequencies of half and one quarter of the maximum. See text for discussion.

```
% scale shown is the frequency, or the square root
% of the frequency, depending on fscale.
% Written by Gerry Middleton, December 1996
if typ == 2     % double, then divide by 2 if necessary
  k = find(az > 179.99);
  az(k) = az(k) - 180;     % corrected az
  az2 = [az;(az+180)];     % extended az
end
x = [0:360/nb:360];
x2 = [180/nb:360/nb:(360 - 180/nb)];   % class limits
if typ == 1
  [f,x2] = hist(az,x2);   % draw histogram
  stairs(x, [f 0]);
else
  xx = [0:360/nb:180];    % change x scale
  x3 = [180/nb:360/nb:180-180/nb];   % frequencies for histogram
  [f,x3] = hist(az,x3);   % draw histogram
  stairs(xx, [f 0]);
  [f, x2] = hist(az2, x2); % frequencies for rose
end
ftot = sum(f);     % find total frequency
if typ == 2
  ftot = ftot/2;   % modify if 180 data
end
fmax = max(f);     % set up axes for histogram
if typ == 1
  axis([0 360 0 (fmax+1)]);
else
  axis([0 180 0 (fmax+1)]);
end
set(gca, 'XTick',x);
figure;            % begin new figure for rose
ff = f;
if fscale == 1   % scale frequency by maximum
  f = sqrt(f);
  fmax = max(f);
  f0 = fmax/10;
else
  f0 = fmax/10;
end
xt = [-fmax:fmax/4:fmax];
yt = xt;
% plot axes
plot([-fmax,-f0],[0,0],'r', [f0 fmax], [0, 0],'r',...
```

```
   [0,0],[-fmax,-f0],'r', [0 0], [f0,fmax], 'r');
centax2;
axis('square');
axis('off');
hold on
% blank out central square, and draw inner circle
h = fill([-f0 -f0 f0 f0],[-f0 f0 f0 -f0],[0.8 0.8 0.8],...
    'EdgeColor',[0.8 0.8 0.8]);
arc(f0,0,360);
text(-0.6*f0, 0, num2str(ftot));  % print total frequency
for i = 1:nb      % draw arcs
   arc(f(i), x(i), x(i+1));
end
rs10 = f0*sin(pi*x(1:nb)/180);   % draw radial lines
rc10 = f0*cos(pi*x(1:nb)/180);
rs1 = f(1:nb) .* sin(pi*x(1:nb)/180);
rc1 = f(1:nb) .* cos(pi*x(1:nb)/180);
rs20 = f0*sin(pi*x(2:nb+1)/180);
rc20 = f0*cos(pi*x(2:nb+1)/180);
rs2 = f(1:nb) .* sin(pi*x(2:nb+1)/180);
rc2 = f(1:nb) .* cos(pi*x(2:nb+1)/180);
k = find(f > 0);
plot([rs10(k); rs1(k)],[rc10(k); rc1(k)],'y','LineWidth',2);
plot([rs20(k); rs2(k)],[rc20(k); rc2(k)],'y','LineWidth',2);
arc2(fmax, 0, 360);  % draw dotted circles at fmax and fmax/2
arc2(fmax/2, 0, 360);
if fscale == 1       % and at fmax/1.4 or fmax/4
    arc2(fmax/sqrt(2), 0, 360)
else
    arc2(fmax/4, 0, 360)
end
hold off

function arc(r, az1, az2);
% arc(r,az1, az2)
% This function draws an arc of a circle
% at a radius r from the origin, from
% az1 to az2 where these are azimuths in
% degrees measured from north
% Written by Gerry Middleton, December 1996
azinc = 5;
az = [az1:azinc:az2];
len = length(az);
rs = zeros(len,1);
```

```
rc = zeros(len,1);
rs = r*sin(pi*az/180);
rc = r*cos(pi*az/180);
plot(rs, rc,'LineWidth',1)
```

Function `arc2` is very similar to `arc` except that it draws a dotted circular arc. The dotted circles in the diagrams indicate the maximum frequency (which is used to scale the diagrams), and frequencies of half and one quarter of that frequency. The central circle is drawn to avoid visually displeasing convergence of radial lines: the central region is also used to display the number of data. Like many graphics programs, this one is more complicated than might be expected, because of the need to allow for several possible options. At least it has the advantage of being easily modified by the user who wants yet another option.

9.2.2 Stereographic Projections

Stereographic projection has been widely used in crystallography, structural geology, and geophysics, for plotting spherical orientations. The use of graphical methods will no doubt decrease as portable computers are available to process data in the field as well as in the laboratory, but stereographic projections still have value as a form of graphic presentation. To write MATLAB functions that produce these nets and plot data on them, it is necessary first to give a brief review of the theory.

Figure 9.2 shows a vertical section through a radius which intersects the lower hemisphere at the point P, and is to be plotted as the projected point P'. If the dip or plunge of the line was originally measured as `dip` in degrees, then it must be converted into `rdip`, measured in radians, using

$$rdip = pi*dip/180$$

Then the angle subtended at the zenith Z is

$$phi = pi/4 - rdip/2$$

If the radius is arbitrarily set to be equal to one, the distance OP' is

$$OP' = \tan(phi)$$

Figure 9.3 shows a plan view of the projection. If the orientation of the line was originally measured as an azimuth `az`, then it must be converted into an angle in radians, `theta` measured anticlockwise from the east

$$theta = pi*(90 - az)/180$$

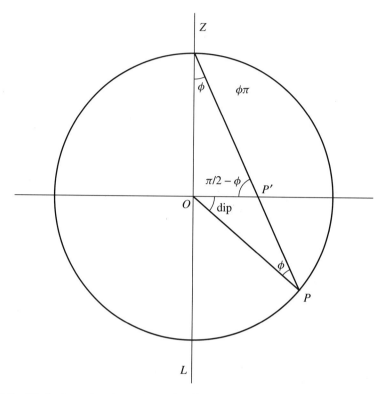

Figure 9.2   Vertical section through a point P to be projected from the lower hemisphere onto the horizontal surface at the point P'

In the case shown, the angle will be $-\texttt{theta}$. The x, y coordinates of P' will be

$$
\begin{aligned}
X &= OP' \cos(\texttt{theta}) \\
Y &= OP' \sin(\texttt{theta})
\end{aligned}
$$

Note that, in the case shown, Y will be negative.

Great circles are plotted by plotting the locus of a radius that lies on a plane of fixed dip, and intersects the lower hemisphere at Q. The projection of this point is Q', given by

$$
OQ' = \tan(\texttt{adip})
$$

where plunge of the radius OQ (the apparent dip of the plane, in the direction of OQ) is

$$
\texttt{adip} = \arctan(\texttt{rdip}) \sin(\texttt{psi})
$$

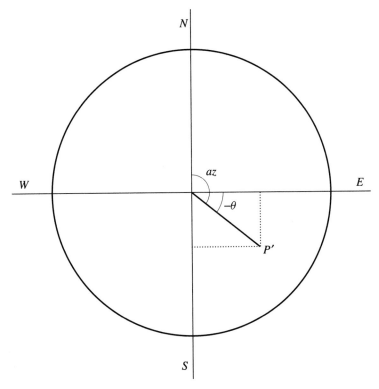

Figure 9.3   Plan view of the projection

The line OQ' makes an angle psi with the north, as shown in Figure 9.4. The coordinates are therefore

$$X = OQ' \sin(\mathtt{psi})$$
$$Y = OQ' \cos(\mathtt{psi})$$

The other set of lines making up the grid on a stereographic projection are the projections of small circles produced by vertical planes that cut the lower hemisphere. On the central horizontal plane, these vertical planes intersect the circular border of the projection at angles of alpha $= \pm 10, 20, ...80$ degrees from the North and South. In the case of the regular stereographic projection, or **Wulff net**, the projections of the small circles are themselves circular arcs, with centers at a distance d from O given by

$$d = 1/\cos(\mathtt{alpha})$$

so they are easily plotted.

For some purposes, we wish to display the density of plotted points on the surface of a sphere. Ordinary stereographic projections are not appropriate for this purpose, because equal areas on the sphere surface do not become equal areas in the projection. To produce

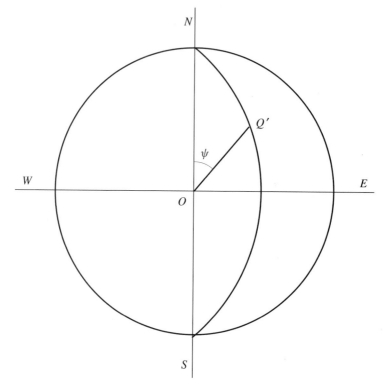

Figure 9.4   Plan of a great circle

an equiarea stereographic projection, or **Schmidt net**, points are projected by circular arc onto a point M on the horizontal plane tangent to the base of the lower hemisphere, before being projected back to the central horizontal plane. This means that the distance OP' is replaced by

$$OM' = \sqrt{2}\sin(\texttt{phi})$$

(see Hobbs et al., 1976, for details).

The small circles are no longer circular arcs, and must be generated by tracing the projection of the small hemicircles. If these intersect the horizontal plane at the angle `alpha0`, then for projections on the horizontal plane of `alpha` ($<$ `alpha0`) the plunge is given by

$$\texttt{adip} = \arccos(\cos(\texttt{alpha0})\sec(\texttt{alpha}))$$

MATLAB provides a `polar` plot which can be used to plot points, but the set of graphic axes (net) that accompany this command is far from being a stereographic net. Therefore the first part of each pair of scripts produces a conventional net, using the regular `plot` command. After this has been run, a file of data may be plotted on the net by first typing `hold` on and then using `wnetplot.m` or `snetplot.m`. Alternatively, data may be plotted directly from the program using the equations indicated in the comments on the scripts.

The following are a series of MATLAB scripts for producing stereonets and plotting files of points on them. The file format for the points must be ASCII, with two real numbers per row: dip (plunge) and azimuth. So if the data are planes, they must be entered as plunges and azimuths of the poles to the planes. The upper diagram in Figure 9.5 shows a Wulff net with paleomagnetic data (fleb2.dat from Fisher et al., 1987, Appendix B2) that display a well defined cluster. The lower diagram in the same figure shows a Schmidt net with structural data (fleb4.dat from Fisher et al., 1987, Appendix B4) that display a well defined girdle.

```
% wulff -- Script for plotting a Wulff net
% to plot points, first calculate
% theta = pi*(90-azimuth)/180 then
% rho = tan(pi*(90-dip)/360), and finally the components
% xp = rho*cos(theta) and yp = rho*sin(theta)
% Written by Gerry Middleton, November 1995
N = 50;
cx = cos(0:pi/N:2*pi);              % points on circle
cy = sin(0:pi/N:2*pi);
xh = [-1 1];                        % horizontal axis
yh = [0 0];
xv = [0 0];                         % vertical axis
yv = [-1 1];
axis([-1 1 -1 1]);
axis('square');
plot(xh,yh,'-g',xv,yv,'-g');        % plot green axes
%axis off;
hold on;
plot(cx,cy,'-w');                   % plot white circle
psi = [0:pi/N:pi];
for i = 1:8                         % plot great circles
   rdip = i*(pi/18);                % at 10 deg intervals
   radip = atan(tan(rdip)*sin(psi));
   rproj = tan((pi/2 - radip)/2);
   x1 = rproj .* sin(psi);
   x2 = rproj .* (-sin(psi));
   y = rproj .* cos(psi);
   plot(x1,y,':r',x2,y,':r');
end
for i = 1:8                         % plot small circles
   alpha = i*(pi/18);
   xlim = sin(alpha);
   ylim = cos(alpha);
   x = [-xlim:0.01:xlim];
   d = 1/cos(alpha);
```

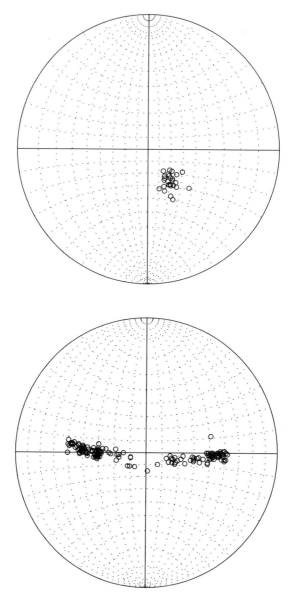

Figure 9.5   Upper diagram shows a Wulff net: the data plotted are magnetic remanences, taken from Fisher et al. (1987, Appendix B2)—they show a well defined cluster. Lower diagram shows a Schmidt (equal area) net: the data shown are facing directions (poles) of conically folded strata, taken from Fisher et al. (1987, Appendix B4)—they show a well defined girdle.

```
    rd = d*sin(alpha);
    y0 = sqrt(rd*rd - (x .* x));
    y1 = d - y0;
    y2 = - d + y0;
    plot(x,y1,':r',x,y2,':r');
end
axis('square');


% Script Wnetplot
% Plots points on a Wulff net created by wulff.m
% Matrix of dips and azimuths is first loaded
% using the function matfile
% linetype must be inside single quotes, e.g. '+y'
% Written by Gerry Middleton, 1995
X = matfile;
theta = pi*(90-X(:,2))/180; % az converted to MATLAB angle
rho = tan(pi*(90-X(:,1))/360); % projected dist. from origin
xp = rho .* cos(theta);
yp = rho .* sin(theta);
i = input('Type in linetype (e.g., +g) for plot: ');
plot(xp,yp,i);


function  X = matfile()
% X = matfile()
% the filename is input from the keyboard
% after loading the matrix is assigned to X
% Written by Gerry Middleton, November 1995
fprintf(' Data file for input must be in the default directory');
t = input('Type in data file name, e.g., dmat.dat - ','s');
    % Use of 's' specifies the input is string, not number
fprintf('\n'); % '\n' specifies a carriage return
fprintf(t);    % Prints the string input
fprintf(' is loaded and assigned to X \n');
               % Adds the phrase 'is loaded'
fprintf('\n');
eval(['load ',t]); % This is equivalent to 'load dmat.dat'
   % The square brackets indicate that this is a single
   % expression
l = length(t); % determine number of characters in filename
n = l-4;    % subtract 4, i.e., .dat
t = t(:,1:n);   % .dat is now removed, leaving the matrix name
eval(['X = ',t]);   % the matrix is assigned to X
return;
```

```
% schmidt -- Script for plotting a Schmidt net
% to plot points, first calculate
% theta = pi*(90-azimuth)/180 then
% rho = sqrt(2)*sin(pi*(90-dip)/360), and finally
% the components xp = rho*cos(theta) and
% yp = rho*sin(theta)
% Written by Gerry Middleton, November 1995
N = 200;
cx = cos(0:pi/N:2*pi);              % points on circle
cy = sin(0:pi/N:2*pi);
xh = [-1 1];                        % horizontal axis
yh = [0 0];
xv = [0 0];                         % vertical axis
yv = [-1 1];
axis([-1 1 -1 1]);
axis('square');
plot(xh,yh,'-g',xv,yv,'-g');        % plot green axes
axis off;
hold on;
plot(cx,cy,'-w');                   % plot white circle
N = 50;
psi = [0:pi/N:pi];
for i = 1:8                         % plot great circles
    rdip = i*(pi/18);               % at 10 deg intervals
    radip = atan(tan(rdip)*sin(psi));
    rproj = sqrt(2)*sin((pi/2 - radip)/2);
    x1 = rproj .* sin(psi);
    x2 = rproj .* (-sin(psi));
    y = rproj .* cos(psi);
    plot(x1,y,':r',x2,y,':r');
end
for i = 1:8                         % plot small circles
    alpha0 = i*(pi/18);
    calph = cos(alpha0);        % angle from N-S
    aset = [-alpha0:0.002:alpha0]; % subdivide arc into alphas
    adip = acos(calph*sec(aset));  % set of plunges
    rho = sqrt(2)*sin(pi/4 - adip/2);
                                    % projected distance from origin
    xp =  rho.*cos(pi/2 - aset);
    y1p = rho.*sin(pi/2 + aset);
    y2p = rho.*sin(3*pi/2 + aset);
    plot(xp,y1p,':g',xp,y2p,':g');
end
axis('square');
```

```
% Script Snetplot
% Plots points on a Schmidt net created by schmidt.m
% Matrix of dips and azimuths is first loaded
% using function matfile
% linetype must be inside single quotes, e.g. '+y'
% Written by Gerry Middleton, November 1995
X = matfile;
theta = pi*(90-X(:,2))/180; % az to MATLAB angle
rho = sqrt(2)*sin(pi*(90-X(:,1))/360);
                        % projected distance from origin
xp = rho .* cos(theta);
yp = rho .* sin(theta);
i = input('Type in linetype (e.g., +y) for plot: ');
plot(xp,yp,i);
```

A more complicated pair of functions, vgcnt3.m, and plotvc.m, not only plot points, but also draw frequency contours.

9.3 Statistics of Directional Data

9.3.1 Summary Statistics

Ordinary means and standard deviations are not good measures of central tendency and dispersion for vectors, because they depend on the choice of origin, which in most cases is arbitrary. Instead, vector mean direction and length are used. The mean length is, of course, inversely related to the dispersion: vectors uniformly distributed on a circle or sphere have zero mean length, whereas N vectors all oriented in exactly the same direction have a mean length of N.

For circular data, the mean vector is found from the azimuths α_i by first calculating the components V, in the N–S direction, and W, in the E–W direction, and summing:

$$V = \sum \cos \alpha_i \tag{9.12}$$

$$W = \sum \sin \alpha_i \tag{9.13}$$

Then the mean azimuth, and vector magnitude (length) are given by

$$A = \arctan(W/V) \tag{9.14}$$

$$R = \sqrt{V^2 + W^2} \tag{9.15}$$

Commonly, the vector magnitude is normalized by dividing by the number of measurements to yield the *consistency* $L = 100(R/N)$ percent. Note that some authors, e.g., Fisher (1993), use the notation $\bar{R}$, instead of L.

For spherical data, a unit vector can best be considered to be represented by its direction cosines $(\lambda_1, \lambda_2, \lambda_3)$. Any two are sufficient to define the orientation, because the sum of squares of the direction cosines is equal to one. A sample of N direction cosines may be written as a matrix λ_{ij} where $i = 1, 2, \ldots, N$ and $j = 1, 2, 3$. A summary description is provided by the sum of squares and cross-products matrix $(b_{jk}) = (\lambda_{ij})'(\lambda_{ik})$. This is a 3×3 symmetrical matrix, which is equivalent to the square of the mean vector length for circular data. A further simplification is achieved by taking the eigenvalues of this matrix: think of this as rotating the ellipse of sample vector frequencies in three-dimensional space until the reference axes coincide with the principal axes of the ellipse. The three eigenvalues, which are generally standardized by making them sum to unity, then provide a convenient summary of the spatial orientation, in much the same way that the principal stresses or strains (stress ellipsoid, or strain ellipsoid) provide a convenient summary of measured stresses or strains. Woodcock (1977) designated the three standardized eigenvalues, ordered by magnitude, as S_1, S_2, S_3. If the orientation is uniform, the three S_i will be equal; if there is a cluster of vectors (linear orientation) then $S_1 > S_2 \approx S_3$; and if there is a girdle (planar orientation) then $S_1 \approx S_2 > S_3$. The S values can be displayed on a plot of S_1/S_2 versus S_2/S_3 (an eigenvalue ratio graph). An example of a cluster and a girdle are shown in Figure 9.5. Figure 9.6 shows an eigenvalue ratio graph, with the girdle from Figure 9.5 plotted on it.

The following MATLAB function calculates the mean and vector strength for circular data (it also calculates the Rayleigh test probability – see next section):

```
function [theta, L, p] = mcirc(az)
% [theta, L, alpha] = mcirc(az)
% calculates the vector statistics for a vector az of
% ungrouped 360 degree data.
% if the data are 180 degree data they must be doubled before
% using this function: after use theta must be halved.
%    theta is the mean vector direction in degrees;
%    R is the vector magnitude;
%    L is the consistency 100*R/N;
%    alpha is the probability (reject null hypothesis if < 0.05)
% Written by Gerry Middleton, 1995.
N = length(az)
x = pi*az/180;   % convert to radians
v = cos(x);
vsum = sum(v);
w = sin(x);
wsum = sum(w);
theta = atan2(wsum, vsum)*180/pi;
R = sqrt(vsum*vsum + wsum*wsum);
L = 100*R/N;
alpha = exp(-R*R/N);
```

The following MATLAB function calculates the direction cosines, mean vector and eigenvalues S_i for a set of spherical data expressed as dip and azimuth values:

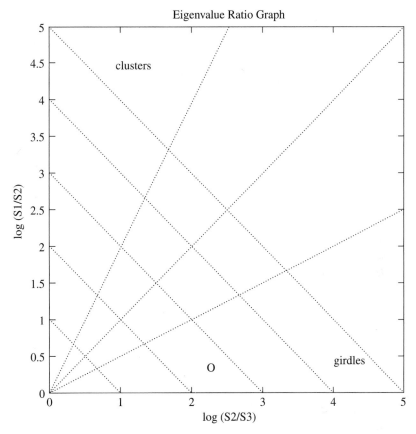

Figure 9.6   Eigenvalue ratio graph: the point plotted is the girdle shown by the data in
the lower part of Figure 9.5.

```
function [dc,dcbar,S] = mspher(dip, az)
% [dc,dcbar,S] = mspher(dip, az)
% converts two column vectors of dips and azimuths
% (in degrees) to a matrix of direction cosines;
% calculates the mean vector, and its strength R;
% and the vector of eigenvalues S, and plots
% the results as an eigenvalue ratio graph.
% Written by Gerry Middleton, December 1996.
N = length(dip);
a = cos(pi*dip/180) .* sin(pi*az/180); % dir cosines
b = cos(pi*dip/180) .* cos(pi*az/180);
c = sin(pi*dip/180);
dc = [a b c];
asum = sum(a);
```

```
bsum = sum(b);
csum = sum(c);
dcbar = [asum bsum csum]/N;    % mean dir cosines
R = sqrt(asum*asum + bsum*bsum + csum*csum);
SS = dc'*dc;
S = eig(SS);
S = -sort(-S); % sort in order of descending magnitude
SR1 = log10(S(1)/S(2));
SR2 = log10(S(2)/S(3));
S = S/sum(S);
plot([0,5],[0,5],':');
axis('square')
hold on;
plot([0,5], [0,2.5],':');
plot([0,2.5], [0,5],':');
plot(SR2, SR1, 'o');
for i = 1:5
   plot([0, i], [i,0],':');
end;
title('Eigenvalue Ratio Graph');
ylabel('log (S1/S2)');
xlabel('log (S2/S3)');
text(1,4.5,'clusters');
text(4,0.5,'girdles');
hold off;
```

The following function performs the reverse function, i.e., it converts direction cosines into dips and azimuths:

```
function v = vec3(dc)
% v = vec3(dc)
% converts a matrix of three columns of direction
% cosines into a matrix of dips and azimuths
dip = 180*acos(dc(3))/pi;
az = 180*atan2(dc(1), dc(2))/pi;
v = [dip az];
```

It should be noted that the direction and magnitude of the mean vector are not always good summary statistics for directional data, any more than the mean and standard deviation are for ordinary data. This is particularly true if the directional data are bi- or polymodal, or show well-developed girdles on stereographic plots.

9.3.2 Statistical Distributions and Tests

Circular and spherical data cannot be Normally distributed, simply because the data are periodic. A variety of other distributions have been proposed as models. For circular data, for example, the von Mises distribution, considered to be a "natural" analog of the Normal distribution, is given by

$$f(x) = \frac{1}{2\pi I_0(\kappa)} \exp(\kappa \cos(x - x_0)) \tag{9.16}$$

where x is a circular variable with mean direction x_0. κ measures the "concentration," i.e., the higher the κ the more preferred the concentration; a uniform distribution has $\kappa = 0$. Cheeney (1983) gives equations for estimating κ from the mean vector length R. I_0 is a rather complicated function of κ. For full descriptions of the model distributions that have been applied to circular and spherical data see Fisher (1993) and Fisher et al. (1987).

Most commonly, we are not interested in the particular distribution, but wish to test whether or not a sample shows evidence of a significant "preferred" orientation, or whether observed differences between the mean vectors of two samples can be considered to be statistically significant. The null hypothesis opposed to a "preferred" orientation is generally the uniform distribution—but different tests evaluate different types of deviation from uniform. A simple Chi Square can be used (see Fisher 1993, p.66–69 for details), but it will test for deviations due to polymodality, as well as due to a single mode. Most researchers want to test the significance of an estimate of the mean vector direction. To do this, they generally use the *Rayleigh* test. For samples of size $N > 50$ calculate

$$\alpha = \exp\left(-\frac{R^2}{N}\right) \tag{9.17}$$

α is the "significance probability" to be compared with the α-level set for the test: so if $\alpha < 0.05$ the null hypothesis of a uniform distribution can be rejected at the 5% level. Equations for smaller samples, and for tests of the difference between a hypothesized mean direction, and an observed mean direction are given by Fisher (1993, Chapter 4). In the older literature one may find references to the "Tukey Chi Square" test: it was a special variant of the Chi Square test that has been shown to be equivalent to the Rayleigh test.

For spherical data, the most common model distribution is the *Fisher distribution*. It is the generalization to a sphere of the von Mises distribution for circular data, originally developed by R.A. Fisher for application to paleomagnetic data (for a review of its history, see Fisher at al., 1987, Chapter 1).

For a simple test of the uniform null hypothesis, the Rayleigh test may still be used. The procedure is to calculate $\chi^2 = 3R^2/N$ and compare with a Chi Squared distribution with three degrees of freedom (Fisher et al., 1987, p.110–111). To assess the difference between two sample mean vectors, or to set up confidence "intervals" for a sample mean, the Fisher distribution is used. The Fisher distribution has three parameters: κ, α, and β. κ is a measure of concentration. If the sample $R/N > 0.95$, then κ is well estimated by $(N - 1)/(N - R)$, otherwise more complicated estimates are required (see Fisher et al., 1987, p.129–130 for details). α and β are estimated by the sample mean spherical angles θ and ϕ (see Section 9.2, Equations 9.4 and 9.5 for calculation of these angles from the mean dip and azimuth). The confidence "interval" in this case is actually a confidence *cone* whose intersection with the surface of the unit sphere plots as a circle. If the sample size is larger than 30, and the estimated $\kappa < 5$, then the semiangle of the cone is given approximately by

$$\theta_\alpha = \arccos[1 + (\log \alpha / \kappa R)] \qquad (9.18)$$

where α in this equation is the confidence level (generally $\alpha = 0.05$). For more details see Cheeney (1983, Chapter 9) or Fisher et al. (1987, Chapter 5).

9.4 Compositional Data

Compositional data are examples of "closed array data." The problem is that, for a set determined on a single specimen, such data must add up to 100 percent. For example, the composition of a rock, consisting entirely of the minerals quartz, feldspar, and mica (a composition closely approximated by some granites, and sandstones derived from them) must satisfy the equation

$$Q + F + M = 100 \qquad (9.19)$$

where Q, F, and M are the percentages of the three minerals. This means that the three compositional variables are not independent, so that there are only two, not three, degrees of freedom. This has the advantage that the composition may be plotted in two dimensions, using a triangular diagram, but it has the disadvantage that it is no longer easy to interpret any trends in composition that might be observed, for example by plotting several samples taken from a single granite mass. Suppose that M remains constant: then there is bound to be a perfect negative correlation between Q and F.

The problem is worse than this: for most rocks the chemical composition can be expressed as a table of about 10–12 chemical oxides: SiO_2, Al_2O_3, TiO_2, and so on. In the past petrologists often showed the variation in oxide abundances by plotting the other oxides against SiO_2 (which is generally the most abundant, and the most variable, oxide percent). Such plots were called *Harker diagrams*. Simply by virtue of the abundance and variability of silica, however, it is expected that most such plots will show a negative correlation between silica and other oxides. Ordinary statistical tests cannot be used to assess the statistical significance of such correlations, because the test calls for independent (and Normally distributed) variables, and so long as the variables are percentages, it is impossible that they be independent (and unlikely that they will be Normally distributed).

The problem extends to ratios. For example, in sedimentology it is common practice to measure three lengths (L, I, and S) to measure the shape of clastic grains, such as pebbles. Shapes can be classified using a plot of I/L versus S/I (a *Zingg diagram*, see Blatt et al., 1980, p.80). But even if there is no statistical correlation between the three measurements, made on a sample of pebbles, there would be a "spurious correlation" between the plotted ratios (for further discussion of the complications in this particular example, see Davis, 1986, p.524–532). Formulas for the spurious correlation induced by taking ratios of originally uncorrelated variables with known variances are given by Chayes (1971).

Petrologists often use ratios of oxides or elements to plot geochemical data, and geomorphologists, sedimentologists and hydraulic engineers use "dimensionless ratios" to plot observations of geomorphological variables or experimental results. For example, the

Froude number $u/\sqrt{gd}$ may be plotted against the Reynolds number ud/ν, where u is the velocity, d is a length scale, g is the acceleration due to gravity, and ν is the kinematic viscosity. Suppose there is relatively little variation in any of the variables, except scale (which, in nature, may vary by several orders of magnitude: for example, river channels vary from less than a meter to more than a kilometer wide). Then a plot of Froude number squared versus Reynolds number is basically a plot of $1/d$ against d, and may be expected to show a strong negative correlation. The correlation itself has no particular physical significance, and tends to obscure other trends between variables that might possibly be physically interesting (for further discussion of an hydraulic example, see Waythomas and Williams, 1988).

Of course, ratio plots may have important advantages that outweigh the problems in statistical interpretation that arise. Scientists who use them, however, should be constantly on guard against misinterpretation of the correlations or "trends" that they show.

9.4.1 Ternary Diagrams

Data consisting of three compositional components (e.g., quartz, feldspar, mica percentages) have something in common with directional data: they can be represented as vectors in three-dimensional space (Philip and Watson, 1988a,b). If the data have been normalized, so that the three components add up to 100 percent, then it will be seen that the data lie on a triangular "simplex," that is, a single triangular plane. This triangle is what is usually displayed as a ternary diagram. Of course, it is not the only way the data could be displayed: the vectors could be normalized to unit length (i.e., the squares of the three components are normalized so that their sum is equal to one), and the data could then be plotted on stereographic projections, and treated statistically as though they were vectors.

Despite their disadvantages, ternary diagrams remain a conventional way to represent compositional data – and there is no built-in MATLAB function that produces this type of plot. The following functions make up for this deficiency.

```
function triplot()
% triplot plots a ternary diagram
% points on it are plotted by tripts
% Written by Gerry Middleton, 1995.
xscale = 2/sqrt(3);
x = [0 xscale/2 xscale 0]; % define coordinates of apices
y = [-1 0 -1 -1];
plot(x,y);
axis('equal');
axis off;    % turn off rectangular axes
hold on;    % hold, to permit adding to plot later
for i = 1:9    % plot dotted lines parallel to sides,
x1 = i*xscale/20;    % at 10 percent intervals
x2 = (20-i)*xscale/20;
x3 = i*xscale/10;
```

```
y1 = -1 + i/10;
y2 = -i/10;
xx = [x1 x2];
yy = [y1 y1];
xx2 = [x1 x3];
yy2 = [y1 -1];
xx3 = [x3 (xscale/2 + x1)];
yy3 = [-1 y2];
plot(xx,yy,'r:',xx2,yy2,'r:',xx3,yy3,'r:');
end

function tripts(X)
% tripts(X) plots points on a triangular diagram
% previously plotted by Triplot.  X is the data matrix,
% with N rows and 3 columns, consisting of x, y, and z:
% the input percentages: the x vertex
% is lower left, y is lower right, and z is top
% Written by Gerry Middleton, 1995
x = X(:,1);
y = X(:,2);
z = X(:,3);
xscale = 2/sqrt(3);
yy = -(100-z)/100;
x3 = x*xscale/100;
x2 = z/173.2;      %tan 60 is 1.732
xx = xscale - x2 - x3;
plot(xx,yy,'+')
text(-0.05,-1,'X');
text(1.2,-1,'Y');
text(0.55,0.05,'Z')
```

Note that the data matrix X must have three columns, and each row must sum to 100. The following function may be used to show the effect of plotting a data set that originally consisted of random numbers, and was then recalculated to percentages.

```
function X = rantri
% function X = rantri
x = 5*rand(50,1);
y = rand(50,1);
z = rand(50,1);
s = x + y + z;
X = 100*[x ./ s, y ./ s, z ./ s];
triplot;
hold on;
tripts(X);
hold off;
```

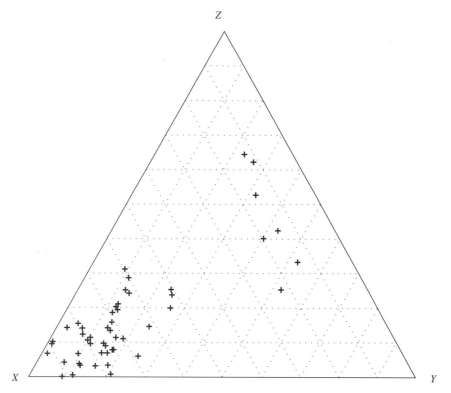

Figure 9.7   Ternary diagram with plot of 50 random numbers. The scale of x was 5 times that of y and z. Before plotting, each set of x, y and z was recalculated to percentages.

The result is shown in Figure 9.7. Note that this plot might give an impression that there was a correlation between x and y, and between x and z.

The purpose of plotting compositional data on ternary diagrams is often to define a field which includes most of the data, for comparison with similar plots of other data sets. For example, sedimentologists wish to plot sandstone compositions in order to compare two different stratigraphic units, or to compare a particular unit with provenance boundaries established by other authors (e.g., Dickinson and Suczek, 1979). The field enclosing most, but not all, the plotted points has been called an *algebraic dispersion field* by Philip et al. (1987), who have described two different ways of defining such a field. Because of the theoretical problems of compositional data, neither can be regarded as true confidence intervals for the data—but they do provide a useful graphical summary of their dispersion. The simpler of the two methods, called the *silhouette method* by Philip et al. (1987), determines the root mean squared difference *(rmsd)* of the mean of the raw compositional data (e.g., percent quartz), then plots a six-sided box around the mean value. The vertices of the box lie at a distance of one *rmsd* from the mean (as normalized to plot in the triangle). The following functions implement the silhouette method in MATLAB.

```
function [m,s] = silh(X)
% [m,s] = silh(X)
% plots the compositional data in X on a ternary
% diagram, determines the mean, m, and root mean
% square of deviations from the mean, s, of the
% values in X, and plots a silhouette diagram
% using the method of Philip et al. (1987)
% X is a matrix with 3 columns (the variables)
% and N rows (the samples)
% Written by Gerry Middleton, Jan. 1997
[N, col] = size(X);
m = mean(X);     % calculate mean
s = std(X);        % and root mean square deviations
    % now calculate vertices of hexagon
S(1,:) = [m(1)+s(1) m(2)-s(2) m(3)-s(3)];
S(2,:) = [m(1)+s(1) m(2)+s(2) m(3)-s(3)];
S(3,:) = [m(1)-s(1) m(2)+s(2) m(3)-s(3)];
S(4,:) = [m(1)-s(1) m(2)+s(2) m(3)+s(3)];
S(5,:) = [m(1)-s(1) m(2)-s(2) m(3)+s(3)];
S(6,:) = [m(1)+s(1) m(2)-s(2) m(3)+s(3)];
sS = sum(S');         % normalize to 100 percent
SS = sS'*ones(1,3);
Sn = 100*S ./ SS;
mn = 100*m/sum(m);
triplot;
hold on;
tripts(X);
tripts2(mn);  % version to plot different symbol
triline(Sn);  % version of tripts to plot polygon
hold off

function triline(X)
% triline(X) plots lines on a ternary diagram
% previously plotted by triplot.  X is a data
% matrix with N rows and 3 columns consisting of
% x, y, and z: the input percentages: the x vertex
% is lower left, y is lower right, and z is top
% The points in X define vertices of a polygon
% written by Gerry Middleton, 1997
[N,col] = size(X);
x = X(:,1);
y = X(:,2);
z = X(:,3);
xscale = 2/sqrt(3);
```

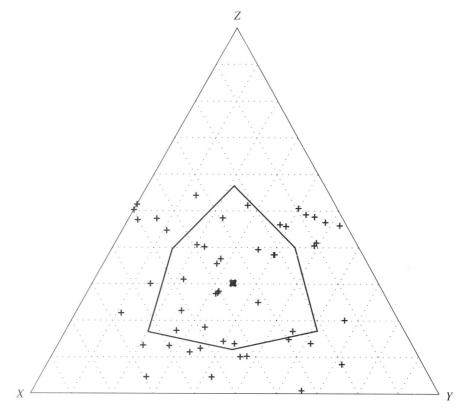

Figure 9.8   Ternary plot of 50 random data, showing mean (heavy symbol) and silhouette dispersion field: the vertices of this field lie one normalized standard deviation above and below the mean values.

```
yy = -(100-z)/100;
x3 = x*xscale/100;
x2 = z/173.2;     %tan 60 is 1.732
xx = xscale - x2 - x3;
yy(N+1) = yy(1);
xx(N+1) = xx(1);
plot(xx,yy,'LineWidth',1);
```

Function `tripts2` differs from `tripts` only in that it plots a different, bolder symbol for the mean point. An example of the use of `silh` is shown in Figure 9.8. For further discussion of the statistics of ternary diagrams see Watson and Philip (1989) and Medak and Cressie (1991).

9.4.2   Conserved Denominator Analysis

One way to reduce, if not eliminate, the effects of closure is to scale the measured variables by taking ratios. For example, weight percent K_2O can vary only from zero to 100 percent, but the ratio K_2O/SiO_2 can (theoretically) have almost any positive value. This type of scaling seems particularly attractive if there is some variable known (or more usually assumed) to be conserved in the process that is under investigation. The advantages then seem to outweigh the possible disadvantages of induced correlation. Among the candidates for conserved denominators are certain "immobile" oxides, rock volume, and conserved elements, including oxygen itself.

One of the earliest applications of this technique was in studies of changes in chemical composition during rock weathering. Goldich (1938) pointed out that, of all the oxides, Al_2O_3 is the least mobile under normal (humid, temperate) conditions. In studying a weathering profile, therefore, one can assume that any apparent increase in alumina is due to loss of more mobile elements such as Na, Ca, Mg, and K. The analyses of weathered rock may be corrected by multiplying them by a factor equal to the ratio of alumina in the unweathered rock to alumina in the weathered rock.

Barth (1948) went further. He strongly advocated using molecular or atomic percentages, and pointed out that, volumetrically, most rocks (and oxide or silicate minerals) are mostly composed of oxygen. Silicate rocks mostly contain about 160 oxygen atoms to 100 cations, so Barth suggested recalculating analyses to a "standard cell" containing 160 oxygen atoms, which is equivalent to using oxygen as the conserved denominator. The following MATLAB function carries out this calculation

```
function [B, P] = barth(X, k)
% [B, P] = barth(X, k)
% calculates a set of chemical rock (or mineral)
% analysis to atoms per unit cell of 160 oxygen
% atoms, following the method described by Barth
% (1948, Jour. Geology, v.56, p.50-61) X is set of
% standard chemical rock analyses with wt % in each
% col, listed in the order: 1=SiO_2,2=TiO_2,3=Al_2O_3
% 4=Fe_2O_3, 5=FeO, 6=MnO,7=MgO,8=CaO,9=Na_2O,
% 10=K_2O, 11=P_2O_3, 12=H_2O+ If any of these are
% omitted, the nos of included oxides must be given in
% the column vector k.  Uses molecular ratios molr.m
% and oxygen scale in os.m (2 for SiO_2, 1.5 for Al_2O_3,
% etc.) B are the cations per 160 oxygens, P are the oxide
% molecular percents. Written by Gerry Middleton, Jan 1997
if nargin < 2
   k = (1:12);
end;
molr;                % load molecular masses
molr = molr(k);      % select appropriate ones
os;                  % load oxygen scale
```

```
os = os(k);             % select appropriate ones
[ox an] = size(X);
for j = 1:an
   p = X(:,j) ./ molr; % scale weights by mol. masses
   s = sum(p);         % recalculate to 100
   P(:,j) = 100*p/s;
   b = P(:,j) .* os;
   s = sum(b);
   B(:,j) = 160*P(:,j)/s;
end
sums = sum(B)
```

The function shown makes use of two M-files that load data into MATLAB. `molr.m` loads the molecular weights of the oxides, divided by 2 if the oxide contains two cations. `os.m` loads the oxygen scale, equal to the number of oxygen to cation atoms in each oxide (so 2 for SiO_2, 1.5 for Al_2O_3, and so on). If one of the standard oxides is omitted from the analyses, this is indicated by the vector `k`. For example, if MnO (the sixth oxide) is omitted, use `k = [1;2;3;4;5;7;8;9;10;11;12]`.

Barth's technique is better suited to studies of mineral than rock compositions, because the stoichiometry of minerals is generally better defined than that of rocks. It was used in a statistical study of scapolites by Middleton (1964).

One problem with most applications to rocks is that there is no assurance that the mass of *any* of the elements has remained unchanged during mass-exchange processes, such as weathering, hydrothermal alteration, or metamorphism. In weathering profiles, for example, alumina is mobilized chemically under extreme chemical weathering, and (in clay) it is moved physically by eluviation (washing) from one level to another. In hydrothermal alteration and metasomatism, there is often textural evidence that, although the mass composition has changed, the rock volume has not. Whitten (1995) has argued from this that analyses should be recalculated to grams per 100 cubic centimeters of rock. Woronov and Love (1990) adopt a more pragmatic approach to identifying conserved elements: they give several examples and suggest how the hypothesis may be tested using logratios (see following).

9.4.3   Pearce Element Ratios

In 1968, Pearce suggested application of a similar technique to the construction and interpretation of variation diagrams for magmatic rocks. His technique has been elaborated by Russell and Stanley (1990) and others. As described for earlier work, the technique relies on using element atomic ratios, rather than weight percents, and the use of models from magmatic petrology, to identify probable conserved (constant) elements, and to construct models based on crystal settling and other mechanisms that alter a parent magma composition. For example, a basaltic magma frequently has its composition changed by crystalization and settling of olivine, $(FeMg)_2SiO_2$, augite, $Ca(MgFeAl)(SiAl)_2O_6$, and/or plagioclase feldspar, $(CaNa)(AlSi_2)Si_2O_8$. Parentheses indicate varying degrees of substi-

tution of one element for another. None of these minerals contain K, which should therefore
be a conserved element, and can be used to form element ratios that will show the true variation much better than simple Harker diagrams (plots of other oxides against SiO_2). The
slope of the lines on such Pearce Element Ratio (PER) plots can also be predicted, using
crystalization models. For example, if only olivine is crystalizing, the composition of the
residual magma should vary along a straight line with a slope of 2 when (Fe+Mg)/K is plotted against Si/K, because olivine contains two atoms of Fe and Mg for every one atom of
Si. The following MATLAB script calculates Pearce element ratios from a set of chemical
analyses (the more complex indices can readily be calculated from the basic ratios returned
by per.m, or the function can be extended to calculate them).

```
function R = per(X, c, k)
% R = per(X, c, k)
% X is a standard chemical rock analysis
% with N columns and 11 rows,
% listed in the order: 1=SiO_2,2=TiO_2,3=Al_20_3
% 4=Fe_20_3, 5=FeO, 6=MnO,7=MgO,8=CaO,9=Na_20,
% 10=K_20, 11=P_20_3
% If any of these are omitted, the list of
% included oxides must be given in the column
% vector k.  c is the number of the conserved
% element to be used in calculating the Pearce
% element ratios returned in R.
% uses atomic ratios in ar.m
% Written by Gerry Middleton, Jan. 1997
if nargin < 3
   k = (1:11);
end;
[ox an] = size(X);
ar;
ar = ar(k);
for j = 1:an         % calculate atomic percentages
   p = X(:,j) .* ar;
   R(:,j) = p/p(c);      % and ratios
end
```

Figure 9.9 shows typical Pearce element ratio plots, for seven hypothetical magmatic
rock analyses (data in file nrex.dat from Russell and Stanley, 1990, p.15–18: this file
excludes oxides 6 and 11). The upper left diagram ((Fe+Mg)/K vs. Si/K) shows two linear
trends. The three analyses with high Si/K ratios show a trend with a slope of 2, which
is consistent with removal of olivine from the original magmatic liquid. The upper right
diagram plots Fc (Fc = 0.25Al + 0.5(Fe+Mg) + 1.5Ca + 2.75Na) against Si/K. There is a
slope of unity, which the authors (Nichols and Russell) interpret as due to the removal of
olivine and/or plagioclase and/or augite (in any combination). However, it might equally
well be due to the addition or removal of almost any elements other than those contained

in these minerals, because of spurious correlation (see Woronow, 1994).

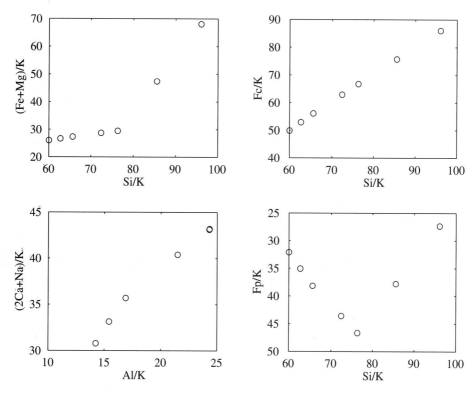

Figure 9.9   Analysis of seven hypothetical rock analyses, using Pearce element ratios (data from Russell and Stanley, 1990, p.15–18). See text for discussion.

The lower left diagram (($2Ca+Na$)/K vs Al/K) is designed to test for removal of plagioclase and/or augite (both minerals contain Ca, Na, and Al) rather than olivine (which does not). The three data points with the highest Al/K ratio all plot at the same position, which seems to confirm that olivine removal is the dominant mechanism producing variation in these rocks (the analyses with the highest Al/K also have the highest Si/K). The slope of one exhibited between the three superimposed points and the next two at lower Al/K ratios are interpreted by the authors as indicating plagioclase removal. They are certainly consistent with this hypothesis, but spurious correlation might also be responsible.

The lower right diagram shows the variation of Fp (Fp = $0.25Al + 0.5(Fe+Mg) - 2.5Ca - 3.25Na$) with Si/K. The diagram shows two very different trends, with slopes of about +1 (at high Si/K ratios) and –1 (at low Si/K ratios). Inspection of the element ratios (tabulated by the authors) shows that, at high Si/K, Al, Ca, and Na ratios are essentially constant, so the trend shown on the right is the same one shown by the top left diagram. At low Si/K ratios, however, the top left diagram shows that the Fe and Mg ratios are almost constant, and the slope of -1 might be due to removal of augite and/or plagioclase, though spurious correlation might also be responsible. For further details and discussion, see Russell and Stanley (1990).

A problem with all methods based on conserved denominator ratios is that this method is particularly likely to produce spurious correlation. In the original data, the values for the conserved element vary. Even if we accept that the chosen element is conserved, the cause of its apparent variation in the original analyses is generally not well understood (indeed, it is the object of investigation). Any good correlation displayed by conserved denominator ratios should be suspect if it has a slope of one—even if it is predicted by a theoretical mixing model. Its existence may be simply the result of plotting a/x versus b/x, where x is variable, and a and b are nearly constant. Woronow (1994) has demonstrated errors of this type in some applications of Pearce element ratios, and suggested an alternative approach based on linear mixing/unmixing models.

9.4.4   Log-ratio Transformation

Another approach to the closure problem, that of using a log transformation of compositional ratios, has been given extensive statistical treatment by Aitchison (1986). The choice of the variable to use in the denominator is arbitrary, but presumably it should be a compositional variable that is determined accurately, yet is not one of primary petrological significance: in a test of Aitchison's method, Rollinson (1992, and see 1993, p.41–42) used TiO_2. At first, it is difficult to see why it is an improvement to use log-ratios (e.g., $\log(SiO_2/TiO_2)$) rather than simple ratios. The reason, according to Aitchison, is that it improves the properties of the covariance matrix, so gives better results for statistical methods that use covariances— such as principal components, discriminant functions, multivariate tests on differences between the mean compositions of different samples, and so on. Aitchison has shown that simple arithmetic means of percentage data have the disadvantage that they may lie *outside* the range of the data, when plotted in multivariate space (e.g., on ternary diagrams). Of course, this is always possible for any variables that show nonlinear covariation, but for compositional data it can often be avoided by the use of means of logratios rather than the original percentages. The logratio means may be transformed back to percentage units by the reverse transformation

$$x_i = 100[r_i, 1]/(\sum_i^{n-1} r_i + 1) \qquad (9.20)$$

Note that if there are n components x_i, there are only $n-1$ logratios r_i.

The following MATLAB function performs the basic data transformations—after which, the usual methods described in earlier chapters may be applied to the transformed data.

```
function [m, S2, T] = lograt(X)
% [m, S2, T] = lograt(X);
% transforms the compositional data in X
% (percentages for m variables on N specimens)
% using Aitchison's log ratio -- the last column
```

```
% of X contains the normalizing percentage
% m is a vector of (n-1) mean logratios, S2 is the
% covariance matrix (n-1 x n-1), and T is the
% transformed data. The program also prints
% the mean values, transformed back to the
% original percentages
% Written by Gerry Middleton
[N,n] = size(X);
T = log(X);                  % calculate logratio data
B = [eye(n-1),-ones(n-1,1)];
T = T*B';
m = mean(T);    % means
S2 = cov(T);    % covariances
mx = 100*[m, 1] / (sum(m) + 1) % means as percentages
```

Most earth scientists have been reluctant to apply a transformation that has little justification derived from scientific (as opposed to statistical) theory. Aitchison and coworkers have also developed several special techniques appropriate to logratio data. The methods need to be tested against well-understood real data; as for example in the study by Rollinson (1992) of fractional crystallization in a Hawaiian lava lake.

We must conclude that, at present, there are no generally accepted techniques for statistical analysis of compositional data. Most earth scientists have learned to live with the problem, and circumvent it, if possible, by the use of calculations based on geologically reasonable models. Lines are fitted to data, and boundaries between data clusters are drawn on graphs (including triangular diagrams)—but few authors aspire to the luxury of statistical tests.

9.5 Recommended Reading

Good introductions to the treatment of directional data are given in several earth science texts:

Cheeney, R.F., 1983, Statistical Methods in Geology. London, George Allen and Unwin, 169 p. (QE33.2.M3C48 See Chapters 8 and 9.)

Davis, John C., 1986, Statistics and Data Analysis in Geology. New York, John Wiley and Sons, Second edition, 646 p. (QE48.8.D38 Directional data are included in Chapter 5.)

Swann, A.R.H. and M. Sandilands, 1995, Introduction to Geological Data Analysis. Oxford, Blackwell Science, 446 p. (QE33.2.S82S93 Chapter 5 deals with directional data.)

More advanced treatment is found in:

Fisher, N.I., 1993, Statistical Analysis of Circular Data. Cambridge University Press, 277 p. (QA276.F488)

Fisher, N.I., T. Lewis and B.J.J. Embleton, 1987, Statistical Analysis of Spherical Data. Cambridge University Press, 329 p. (QA276.F489 1987)

Many books and papers discuss the use of stereographic projections, but few give details about how the nets used are constructed. The following are exceptions.

Hobbs, B.E., W.D. Means and P.F. Williams, 1976, An Outline of Structural Geology. New York, John Wiley and Sons, 571 p. (QE601.H6 see Appendix A)

Ragan, D.M., 1985, Structural Geology: An Introduction to Geometrical Techniques. New York, John Wiley and Sons, Third Edition, 393 p. (QE601.R23)

Spencer, A.B. and P.S. Clabaugh, 1967, Computer programs for fabric diagrams. American Journal of Science, v.265, p.166–172.

Suppe, John, 1985, Principles of Structural Geology. Englewood Cliffs NJ, Prentice-Hall, 537 p. (QE601.S94)

A large number of computer programs for plotting nets, plotting on nets, and contouring data plotted on nets have also been published. The following is one of the more recent:

Van Everdingen, D.A., J.A.M. Van Gool and R.L.M. Vissers, 1992, Quickplot: a micro-computer-based program for the processing of orientation data. Computers and Geosciences, v.18, p.283–287.

Closed data and ratio correlation are discussed by:

Aitchison, J., 1984, Statistical analysis of geochemical compositions. Mathematical Geology, v.16, p.531–564. (A better introduction than the next reference.)

Aitchison, J., 1986, The Statistical Analysis of Compositional Data. London, Chapman and Hall, 416 p. (QA278.A37 Monographic treatment.)

Chayes, Felix, 1971, Ratio Correlation. University of Chicago Press, 99 p.

Medak, F. and N. Cressie, 1991, Confidence regions in ternary diagrams based on the power-divergence statistics. Mathematical Geology, v.23, p.1045–1055.

Philip, G.M., C.G. Skilbeck, and D.F. Watson, 1987, Algebraic dispersion fields on ternary diagrams. Mathematical Geology, v.19, p.171–181.

Philip, G.M. and D.F. Watson, 1988a, Determining the representative composition of a set of sandstone samples. Geological Magazine, v.125, p.267–272.

Philip, G.M. and D.F. Watson, 1988b, Angles measure compositional differences. Geology, v.16, p.976–979.

Reyment, R.A., 1989, Compositional data analysis. Terra Nova, v.1, p.29–34 (A brief review.)

Rollinson, H.R., 1992, Another look at the constant sum problem in geochemistry. Mineralogical Magazine, v.47, p.267–280.

Rollinson, H.R., 1993, Using Geochemical Data: Evaluation, Presentation, Interpretation. Harlow, Essex (England), Longman, 352 p. (QE515.R75, Chapter 2 has a good discussion of compositional data.)

Russell, J.K. and C.R. Stanley, eds., 1990, Theory and Application of Pearce Element Ratios to Geochemical Data Analysis. Geological Association of Canada Short Course v.8, 315 p. (QE515.T49)

Watson, D.F. and G.M. Philip, 1989, Measures of variability for geologic data. Mathematical Geology, v.21, p.233–254.

Waythomas, C.F. and G.P. Williams, 1988, Sediment yield and spurious correlation—towards a better portrayal of the annual suspended-sediment load of rivers. Geomorphology, v.1, p.309–316.

Whitten, E.H.T., 1995, Open and closed compositional data in petrology. Mathematical Geology, v.27, p.789–806.

Woronow, Alex, 1994, Identifying minerals controlling the chemical evolution of igneous rocks: Beyond Pearce element-ratio diagrams. Geochemica et Cosmochimica Acta, v.58, p.5479–5487.

Woronov, Alex and K.M Love, 1990, Quantifying and testing differences among means of compositional data suites. Mathematical Geology, v.22, p.837–852.

Chapter 10

Fractals

10.1 Introduction

Fractals are geometric objects, therefore they are defined mathematically. The term was first published in 1975 by Benoit Mandelbrot, in the first, French, edition of the book whose latest English edition was published in 1982. Unfortunately, there is still no rigorous definition of the term that is universally accepted: the simplest, given by Feder (1988, p.11) and attributed by him to Mandelbrot, is

> A fractal is a shape made of parts similar to the whole in some way.

In the latest, French edition of his book, Mandelbrot (1989, p.154) expands on the meaning of self-similarity by offering the following "intuitive" definition (translated by the present author):

> A geometric figure or natural object is said to be fractal if it combines the following characteristics: (a) its parts have the same form or structure as the whole, except that they are at a different scale and may be slightly deformed; (b) its form is extremely irregular, or extremely interrupted or fragmented, and remains so, whatever the scale of examination; (c) it contains "distinct elements," whose scales are very varied and cover a large range.

The self-similarity characteristic of fractals may be exact, or it may be only statistical in nature. Most natural fractal objects show only statistical self-similarity, but there are some exceptions (e.g., plants: see Peitgen et al., 1992, Chapter 5). As indicated by Mandelbrot's oblique reference to "slightly deformed" parts, there are also fractals that are not quite self-similar, in that they scale differently in different directions (e.g., in the horizontal and vertical directions): they are called *self-affine*.

An example of a fractal that shows exact self similarity is the *Koch curve*. It is generated from a unit line by replacing the mid one third with two line segments, each one third of the original length, as shown in Figure 10.1, the upper two diagrams. Thus a single line of unit

194

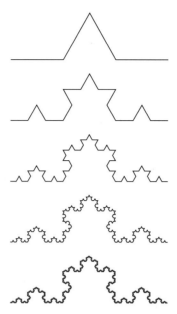

Figure 10.1   Generation of the Koch curve: the top diagram (at stage 0) is the *initiator*,
and the next (at stage 1) is the *generator*.

length is replaced by four line segments, each of length 1/3. The process is then repeated
ad infinitem: at each stage, each line segment is replaced by four segments, and the total
length increases by a factor of 4/3. It can be seen that the first third of the curve at stage
2 is exactly similar to stage 1, but at one third of the scale. Enlarging the first third of the
"finished" curve by three times would exactly reproduce the whole curve. The following
function generates the Koch curve, up to stage n.

```
function X = koch(n)
% X = koch(n)
% generates a koch curve up to the nth stage
% X contains the coordinates of all points
% on the curve.  The beginning and end points
% are (0,0) and (1,0); columns are x and y values,
% rows are data points, so after the first stage
% of construction there are 5 points.  At each
% stage the number of points N(i) is 4*N(i-1) - 3
% or 4^n + 1.  So n should not be set larger than
% 7 (16,385 points!).
% Written by Gerry Middleton, Jan. 1997
if n > 7
   error('n should be < 8');
end;
```

```
X = zeros(4^n+1,2);    % reserve space
XX = zeros(4^n+1,2);
X(1,:) = [0 0];        % stage zero
X(2,:) = [1 0];
subplot(2,1,1);
plot(X(1:2,1), X(1:2,2));
axis([0 1 0 0.3]);
axis('equal');
subplot(2,1,2);        % stage 1
XX(1,:) = [0 0];
XX(2,:) = [1/3 0];
XX(3,:) = [1/2 sqrt(1/12)];
XX(4,:) = [2/3 0];
XX(5,:) = [1 0];
plot(XX(1:5,1),XX(1:5,2));
axis([0 1 0 0.3]);
axis('equal');
N = 5;
X = XX;
m = 1;
for i = 2:n             % stages 2 on up
   k = 1;
   for j = 1:N-1
      dx0 = X(j+1,1) - X(j,1);
      dy0 = X(j+1,2) - X(j,2);
      h = sqrt(1/(4*3^(2*i-1)));  % distance from apex to base
      theta = atan2(dy0, dx0);
      dx = h*sin(theta);
      dy = h*cos(theta);
      XX(k+1,:) = [X(j,1)+dx0/3, X(j,2)+dy0/3];
      XX(k+2,:) = [X(j,1)+dx0/2-dx, X(j,2)+dy0/2+dy];
      XX(k+3,:) = [X(j,1)+2*dx0/3, X(j,2)+2*dy0/3];
      XX(k+4,:) = X(j+1,:);
      k = k+4;
   end
   if i < 6  % plot only stages up to 5
      if m == 2
         m = 1;
         figure;
         subplot(2,1,1);
         plot(X(1:N,1), X(1:N,2));
         axis([0 1 0 0.3]);
         axis('equal');
         subplot(2,1,2);
```

```
          plot(XX(1:4*N-3,1),XX(1:4*N-3,2));
          axis([0 1 0 0.3]);
          axis('equal');
      else
          m = m+1;
      end;
   end;
   X = XX;
   N = 4*N - 3;
end;
```

A version of the Koch curve that shows only statistical self-similarity can be produced by using more than one generator, selected at random for each line segment (see Hastings and Sugihara, 1993, p.26–28 for an example). A better example of a fractal that shows statistical self-similarity is the curve produced by a random walk. The original Brownian motion, produced by molecules in solution colliding with tiny dust particles is such a curve in three dimensions. The two-dimensional version, illustrated in many texts (e.g., Peitgen et al., 1992, p.490) shows self-similarity. Random walks in one dimension are generally displayed in two dimensions; the x-coordinate is incremented regularly at each step, but the y-coordinate is formed from the sum of a series of increments Δy, randomly distributed about zero. These random walks are self-affine rather than self-similar, and can be used as models of diffusion, as illustrated in the following MATLAB script.

```
% Script diffuse2.m
% Plots 500 random walks, each of length
% 1000 steps, and records total displacement
% in y direction. Two plots are shown:
% the first is arithmetic, the second plots
% the square of displacement against steps.
% After walks are plotted a histogram of the
% displacements is shown.
% Have patience, computation takes a while!
% Press <space> to begin, and to see last screen
% Written by Gerry Middleton, 1996.
help diffuse2;
pause;
x = [1:1000]';
f = zeros(500,1);
figure;
for i = 1:500
   y = rand(1000,1) - 0.5;
   yy = cumsum(y);
   if rem(i,10) < 0.001
      subplot(211);
      plot(x,yy), axis([1 1000 -25 25]);
```

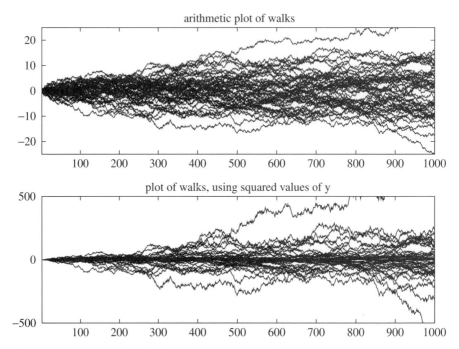

Figure 10.2   A series of random walks, beginning from the same point, showing the
resulting pattern of diffusion down-walk.

```
    if i == 10
        hold on;
        title('arithmetic plot of walks');
    end
end
if rem(i,10) < 0.001
    subplot(212)
    yy2 = abs(yy) .* yy;
    plot(x,yy2), axis([1 1000 -500 500]);
    if i == 10
        hold on;
        title('plot of walks, using squared values of y');
    end
end
f(i) = yy(1000);
end
pause;
figure;
hist(f), title('histogram of 500 Delta y at end of walk')
```

To produce an enlarged version of a part of this kind of random walk, which looks (statistically) just like the original, it is necessary to enlarge the x and y dimensions by different factors: such fractals are described as self-affine rather than self-similar. It can be shown that the average distance away from the origin is proportional, for long walks, to the square root of the number of steps. This effect is demonstrated by the script `diffuse2` by squaring the y scale in the second plot, thus producing a more linear "cone of diffusion." To enlarge a portion of a random walk, therefore, if the x dimension is scaled by k, the y dimension must be scaled by $k^{0.5}$.

10.2   Power-law Scaling

In the earth sciences, one of the best known examples of power-law scaling is the Gutenberg-Richter law. For a given region, and over a given period of observation, it has been found that the number of earthquakes N with a magnitude greater than m are related by

$$\log N = -bm + \log a \qquad (10.1)$$

where b is almost everywhere in the range $0.8 < b < 1.2$ and a is a coefficient that varies from one region to another (it is larger in regions of higher seismicity). The magnitude is proportional to the logarithm of the energy of the earthquake.

Consider some of the consequences, if we accept this law at its face value. The fact that the *scaling exponent b* is a constant means that the scaling of earthquakes does not depend on their magnitude: whatever the scale, earthquakes one magnitude weaker are likely to be about ten times as abundant. This suggests some sort of self-similarity in the mechanics of earthquakes. We will see in the next section that fractal objects display comparable scaling laws. The existence of power-law scaling is therefore cited as evidence for the fractal nature of natural phenomena. For description of many other examples in the earth sciences, see Turcotte (1992, 1994) and Thompson (1991).

Does power-law scaling prove self similarity? The following example shows that power-law scaling can be shown by objects not generally considered to be self similar. Consider a set of points uniformly distributed in the plane, so the number of points in a region is proportional to its area. Then the number of points located with a distance r of some reference point is proportional to r^b, where $b = 2$. This is a power law of the same type as Equation 10.1, but uniformly distributed points are not generally considered fractal. One might object that the exponent is an integer, and if it were not, the distribution of points might be considered fractal. An example with this property would be a cloud of points, whose density decreases as some function of the distance r from its center. In this case the exponent b would not be equal to 2, and might well be fractional. Some writers have used such observations to establish, for example, that the distribution of oil fields is fractal, but to do so seems to be stretching "statistical self-similarity" to the point where it become meaningless.

An important consequence of power-law scaling is that a population showing such scaling has no characteristic average value. If the Gutenberg-Richter law is strictly true, there is no average magnitude of earthquake. One can speak only of the distribution of sizes, for example, of the probability that an earthquake exceeding a given size will occur within the next 10, or 100, years. In nature, however, power-law scaling is generally observed only over a limited range of scales. An example is the size and strength of turbulent eddies, as recorded, for example, by the strength and duration of velocity fluctuations. In a river, the largest scale is limited by the size (depth, width) of the river, and the smallest scale is limited by the effects of viscosity. So even though the total energy of eddies shows power-law scaling with scale (or frequency) of eddies, it does make sense to calculate an average velocity (or kinetic energy) for a river.

10.3   Fractal Dimension

If, in theory, one cannot determine the average size of a fractal object, how can it be characterized? The answer is, by its *fractal dimension*. Unfortunately, it is not easy to give a rigorous definition of geometric dimension. The one generally cited (the Hausdorff-Besicovich dimension) is not only difficult to understand, but is almost never used in practice. In practice, fractal dimensions are determined by measuring scaling dimensions.

For a geometrically-constructed object, like the Koch curve, it is not difficult to determine the fractal dimension. The fractal dimension is determined from the scaling equation relating the length of the curve L to the size δ of the object used to measure it

$$L = a\delta^{1-D} \tag{10.2}$$

At the pth stage of construction, the δ may be taken to be the length of the line segment, given by

$$\delta = \left(\frac{1}{3}\right)^p \tag{10.3}$$

and the total length of the Koch curve is

$$L = \left(\frac{4}{3}\right)^p \tag{10.4}$$

so the scaling equation must satisfy

$$(4/3)^p = [(1/3)^p]^{-(D-1)} \tag{10.5}$$

Taking logarithms of both sides gives the result that

$$D = \log 4 / \log 3 \approx 1.26\ldots \tag{10.6}$$

For a natural curve Equation 10.2 may be used to determine the fractal dimension by using as the measuring object either a pair of dividers, with spacing δ, or a set of covering boxes of side δ. The classic example, originally investigated by L.F. Richardson, and republished by Mandelbrot (1967) is the determination of the length of coastlines by the dividers method. The dividers method is not particularly easy to implement by computer: various suggested techniques have been discussed and compared by Andrle (1992, 1996) and Brown (1995). The first script given here is designed for curves whose digitized coordinates are given in a matrix **X**: it is based on the technique described by Andrle. The algorithms used are designed to avoid program loops as far as possible, because these execute slowly in MATLAB. The basic method used to determine the number of strides, of length d, necessary to cover the line, is as follows:

1. select a starting point;

2. select from the data a submatrix that almost certainly includes points more distant than d;

3. find the two points at distances that bracket d;

4. interpolate between those two points (`nextpt` and `nextpt2`) or chose the closest point (`nextpt3`);

5. this point is now the new starting point, and the search continues for the next point;

6. close to the end, the method changes to permit calculation of a final fractional stride.

`divplot` simply uses the first point as the starting point. The value of d is initially set equal to d_{min}, which has a value of twice the maximum distance between successive points in **X**. After this d is set to higher powers of d_{min}, until the number of strides necessary to cover the data is less than 10.

`divplot2` chooses 10 initial points at random, and strides from that point both up and down the data matrix. `steps` is the function that calculates the number of strides. Its main task is item (2) in the list given above. `steps` and `steps2` are 'stand-alone,' 'demo' programs that accept a data matrix and a value of d, and plots the result. The plot will generally only be interesting for values of d that are a large fraction of the overall dimension of the data set.

`steps3` is designed to be called by `divplot2`: it returns only the number of strides and omits the plot. The task of determining the next point at each stride is assigned to `nextpt`. Three varieties are given:

- `nextpt` implements an approximate interpolation technique. After locating the two points that are just more and less distant than d (`d1(1)` and `d1(2)`), it uses the ratio of differences in distance from d, to interpolate between the two points. The formula used is to interpolate between two points, at a ratio λ (for derivation, see Gellert et al., 1977, p.286-288). It is a good approximation only when the angle that the two points subtend from the initial point is small, which is generally but not invariably true. The distance to the interpolated point is always less than the true distance d.

- `nextpt2` is a closer approximation to the "exact" interpolation which increments the estimated lambda until the distance lies very close to the true distance. It is considerably slower, because it makes use of a 'while' loop, which executes slowly in MATLAB.

- nextpt3 is the "fast" method, which simply selects as the next point, the point whose distance is closest to d.

The slopes provided by divplot are those obtained by fitting a straight line to all the data. The fractal dimension D is equal to one minus the slope. It is essential to check the graph to determine if the line is, in fact, a good fit. If it is not, then the curve does not show true fractal properties, and the dimension determined is essentially meaningless.

```
function [d,dn] = divplot2(X)
% [d,dn] = divplot2(X)
% plots a divider plot, log(dn) vs log(d)
% for divider lengths, d, that are powers of 2 of dmin
% where dmin is twice the maximum distance between
% successive points; dn is the total measured distance
% for each divider length d. Calls steps3.m
% This version divplot2 uses a random row in the matrix
% X as starting point, and computes dn both above and
% below this row: the sum of the two is the value returned.
% This is repeated for 10 different starting points.
% Written by Gerry Middleton, July 1996
[r,c] = size(X);
if c ~= 2
   error('Not a list of (x,y) pts')
end;
dmax = max((X(1:r-1,1)-X(2:r,1)).^2 + (X(1:r-1,2)-X(2:r,2)).^2);
dmin = sqrt(ceil(2*dmax));
fprintf('Computing -- please wait');
for j = 1:10
   nn = 100;
   k = rand(1);
   k = round(k*r)
   if k == 0
      k = 1
   end;
   X1 = X(k:-1:1,:);
   X2 = X(k:r,:);
   i = 1
   d(i,j) = dmin;
   while nn > 10
      if k > 1
         nn1 = steps3(X1,d(i,j));
      end;
```

```
            nn2 = steps3(X2,d(i,j));
            nn = nn1 + nn2;
            dn(i,j) = nn*d(i,j);
            i = i + 1
            d(i,j) = 2*d(i-1,j);
        end;
    end;
    d = d(1:i-1,:);
    Ld = log2(d);
    Ldn = log2(dn);
    [r2,c2] = size(dn);
    ldn = reshape(Ldn,r2*c2,1);
    ld = reshape(Ld,r2*c2,1);
    p = polyfit(ld,ldn,1);   % linear regression
    f = polyval(p,ld);
    s = num2str(p(1));
    ss = ['Slope = ' s];
    plot(ld,ldn,'o',ld,f), axis('equal');
    ylabel('log2(total length)'), xlabel('log2(interval)')
    title(ss);

    function n = steps3(X,d)
    % function n = steps3(X, d)
    % calculates the number n of steps to walk along
    % the set of points X using a divider length d
    % calls nextpt2.m  To increase speed at the risk
    % of accuracy change this to nextpt.
    % Written by Gerry Middleton, June 1996
    [r c] = size(X);
    if c ~= 2 error('Not a list of (x,y) pts')
       end;
    i = 1;
    n = 0;
    xn = [X(1,1) X(1,2)]; % first point
    Y(1,:) = xn;
    %
    % now calculate an estimate of the number of points
    % spanning a distance always larger than d
    % Note that this may have to be increased for
    % some data sets by changing the coefficient 5
    %
    Xlag = X(2:r,:);
    delta = (X(1:r-1,1)-Xlag(:,1)).^2 + (X(1:r-1,2)-Xlag(:,2)).^2;
    tdist = sum(delta);
```

```
dbar = mean(delta);
npts = 10*ceil(d/sqrt(dbar));
%
% calculate vectors of distances squared
% and find the next point in each, then move the
% vector along the list of points
%
while i+npts < r
   Xp = [xn; X(i:i+npts,:)];
   [xn, k] = nextpt3(Xp,d);
   n = n + 1;
   i = i+k-2;
end;
%
% calculate the remaining distance squared and
% while it is less than d^2, determine remaining points
%
dm2 = (xn(1) - X(r,1))^2 + (xn(2) - X(r,2))^2;
while dm2>d^2
   Xp = [xn; X(i:r,:)];
   [xn, k] = nextpt3(Xp,d);
   n = n+1;
   i = i+k-2;
   dm2 = (xn(1) - X(r,1))^2 + (xn(2) - X(r,2))^2;
end;
%
% add a fraction at the end
%
dn = sqrt(dm2/d^2);
n = n+dn;


function [xn,k] = nextpt2(X, d)
% [xn,k] = nextpt2(X, d)
% given an r x 2 matrix X containing the coordinates of
% r points as [x,y] pairs, finds the kth point, just
% further than d, and interpolates the point with
% coordinates xn(x,y) which is a distance d from X(1,1).
% Written by Gerry Middleton, June 1996
[r,c] = size(X);
if c ~= 2 error('X not a list of 2D pts')
   end;
d2 = (X(1,1) - X(:,1)).^2 + (X(1,2) - X(:,2)).^2;
   % vector of distances squared
k = min(find(d2>d^2));  % locate distance just larger than d
```

```
d1 = sqrt([d2(k-1) d2(k)]);
  % distances to points just short of and beyond d
dd = d^2;
deldd = dd/1000;
if abs(d - d1(1)) < 0.001*d
   xn = X(k-1,:);
else
   lam = (d - d1(1))/(d1(2) - d);
   % lam{da} is the first approx. of ratio for interpolation
   xn = (X(k-1,:) + lam*X(k,:))/(1+lam);
   d3 = (X(1,1) - xn(1,1))^2 + (X(1,2) - xn(1,2))^2;
   %
   % in general d3 is less than dd
   % if d3 is not close enough, increment lam
   % 20 loops should be enough
   %
   n = 1;
   while ((dd - d3)>deldd) & (n<20)
      lam = dd*lam/d3;
      n = n+1;
      xn = (X(k-1,:) + lam*X(k,:))/(1+lam);
      d3 = (X(1,1) - xn(1,1))^2 + (X(1,2) - xn(1,2))^2;
   end;
end;
```

The result of running `divplot2` on the output from `koch` (with $n = 5$, saved as file `koch5.dat`, and scaled by 100) is shown in Figure 10.3. The slope b is related to the fractal dimension by $D = 1 - b$. The slope is -0.27, so the fractal dimension is 1.27, which differs somewhat from the theoretical value of 1.26. It is disturbing that the plotted points do not seem to lie on a straight line. One might think that this was a result of errors in programming, if it were not consistent with the results of numerical experiments reported by many other investigators. The fact is that it is difficult to obtain numerical estimates of fractal dimensions that are accurate to more than two significant figures. One reason is that, theoretically, the dividers or box-counting methods only give accurate results as δ tends to zero—which is a limit that, for almost all real sets of data, cannot be closely approached.

The next function `divider2` is designed to implement the divider method for random walks, $x = f(t)$, or traverses across surfaces. The idea is to substitute the "time" interval, and various multiples of it, for the divider spacing, and the absolute different between the values of x at this step and the previous step ($\Delta x = \text{abs}(x(t) - x(t + k\delta t))$) for the "length" of the step. I believe this method is equivalent to the more complicated one discussed by Brown (1995). He shows that it should give the correct result if the vertical scale is large enough: to test this, enlarge the vertical scale and see if it gives a similar result.

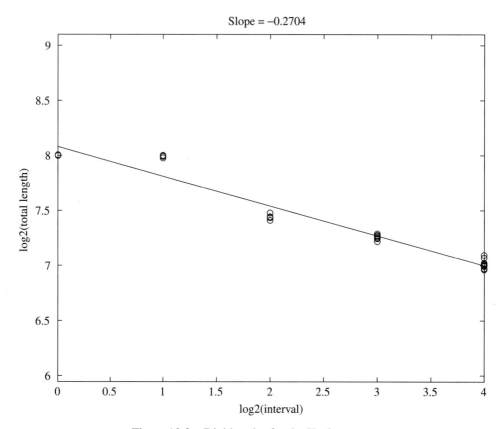

Figure 10.3   Divider plot for the Koch curve

```
function Y = divider2(x)
% Y = divider2(x)
% Plots the total length of a time series x_i, as a
% function of the spacing ni between the values.   x
% is a vector of length N, so i = 1,2,...N. The
% spacing is allowed to increase from 1,2,4,...2^(n-1)
% where 2^n < N/2.  Y has n cols, each of which is the
% total length for a spacing ni, obtained by traversing
% 2^(n-1) points. There are also m rows, each corresponding
% to starting the measurement at 1,2,...m  In this version
% m = 20.  The plot shows m points for each value of the
% spacing, and the regression line is based upon all the
% data except the largest spacing
% Written by Gerry Middleton, Jan. 1996
N = length(x)
```

```
N2 = floor(N/2);
n = floor(log2(N2))
m = 20;
for j = 1:m
   for i = 1:n
      ni = 2^(i-1);      % (i.e., 1,2,4,...)
      nn(i) = ni; % store for plot as row vector
      xi = x(j+ni:ni:j-1+N2+ni);
         % xi is x lagged by ni and j, at intervals of ni
      Y(i,j) = sum(abs(x(j:ni:j-1+N2) - xi));
      end
   end
lnn = log2(nn)';      % log2 of n as column vector
lY = log2(Y); % log2 of Y matrix
for j = 1:m
   plot(lnn, lY(:,j), 'o');
hold on;
end
lYv = reshape(lY(1:n-1,:),m*(n-1),1);
   % remove nth value and reshape
lnv = reshape(lnn(1:n-1)*ones(1,m),m*(n-1),1);
p = polyfit(lnv,lYv,1)  % linear regression
f = polyval(p,lnv(1:n-1));
s = num2str(p(1));
ss = ['Slope = ' s];
plot(lnv(1:n-1),f);
ylabel('log2(total length)'), xlabel('log2(interval)')
title(ss);
```

The result of applying this to a random walk is shown in Figure 10.4. The data file (brown.dat) consisted of 1025 points generated by the mid-point displacement method (implemented in mpdwalk, see Peitgen et al., 1992, p.487–490). Theoretically, the fractal dimension should be 1.50: the value determined by the dividers method is 1.514.

Another commonly used method to determine fractal dimension is *box counting*. In this method, the geometric object is enclosed within a box, which is then subdivided into 4 equal-size boxes. The boxes that contain some part of the object are then counted. The process is repeated (each box being subdivided into 4 at each step) until some minimum size of box is reached. A log-log plot of the number of filled boxes, times their dimension, versus the size of the box, should be a straight line, with a negative slope equal to $(1 - D)$ where D is the fractal dimension (see Feder, 1988, Chapter 2 for further discussion and an example). This method is implemented in the function boxcnt.m (not listed here, but included in the diskette). Applied to the Koch data in koch5.dat, the method yields a somewhat better straight line, and a fractal dimension of about 1.29. The method can also be applied to map data, or two-dimensional random walks.

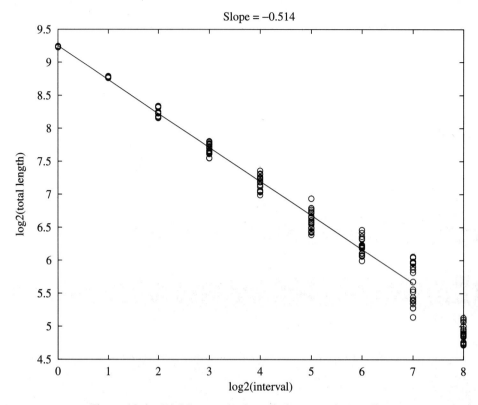

Figure 10.4   Dividers method applied to a random walk.

The literature on practical determination of fractal dimension is now large, and expanding rapidly. We have concentrated on variants of the dividers method, but there are many other methods (e.g., Peitgen et al.,1992; Middleton et al., 1995; Russ, 1994; Klinkenberg, 1994). It is important that investigators realize that some methods may be appropriate only to self-similar rather than self-affine data, and that there are often large differences in fractal dimension estimates obtained by using different methods. The problems arise for several reasons:

- Most natural "fractals" are, at best, fractal over only a limited scale range. Unfortunately, many books and papers present a few power-law plots, with lines drawn by hand or fitted by ordinary least squares, and claim to have "proved" that the objects are fractal despite obvious departures from power-law behavior. Few authors apply goodness of fit techniques, or compare the power-law with alternative distributions.

- Though we too have followed the usual procedure, and used least squares to fit lines to power-law plots, it is clear that this is at best a dubious use of the technique, because it is not clear what weights should be assigned to the points plotted. Plots such as Figure 10.4 clearly indicate that the error variance varies with the independent variable ("heteroscedascity"): a condition that invalidates most techniques for setting confidence intervals on the slope.

- As remarked earlier, theoretically most dimension determination depends on taking a limit as the measuring scale (δ) approaches zero—but there are often good reasons why this limit cannot be approached in the real world. For example, most geomorphic forms are nearly smooth at small scales: coastlines are smooth where there are beaches, and even rocky shores are smooth at about the meter scale. The theoretical fractal dimension of a smooth curve is always equal to one, no matter how convolute it may appear at larger scales. The real problem is to determine whether or not there exists a true power-law scaling range, and if so, what the fractal dimension would be if this scaling extended to infinitesimally small scales (which we know it does not).

- The statistics of fractal distributions have only recently been investigated (e.g., Cutler, 1994), and the few results obtained have not yet been adequately tested on real data.

10.4 Models that Generate Fractals

We have seen a few examples of geometric constructions that generate fractals, but what produces fractal objects in the real world? There is no single answer to this question, just as there is no single answer to the question, what produces circles and ellipses in the real world? The following are a few of the possible answers:

- Fractals can be produced by stochastic processes. One class consists of random (and quasi-random) walks. These include not only Brownian motions, with fractal dimensions of 1.5, but *fractional Brownian motions* (sometimes called pink or black motions) with fractal dimensions anywhere in the range from 1 to 2. These types of fractals have become popular as a way of producing not-quite realistic simulations of topography—simulations that Mandelbrot has called "fractal forgeries."

  Another class includes models of fragmentation that produce fractal size distributions (Turcotte, 1992, Chapter 3). An initial block is broken into eight (or some randomly chosen number) of pieces, each of which has a certain probability of being broken again, and so on. Also in this class are certain "multiplicative cascade" models of turbulence: large eddies break down into a number of smaller eddies, with energy being randomly partitioned between them, and each eddy has a certain probability of breaking down into smaller eddies, and so on. Such models can lead to fractals that can be characterized not only by a geometric fractal dimension, but also by a range of other scaling exponents (*multifractals*).

- Fractals can be produced by nonlinear dynamical systems, that is by physical or biological systems that are controlled by sets of nonlinear differential equations. Many of these systems have a deceptive simplicity, being controlled by only three or more equations. The behavior of such systems may be perfectly regular and predictable in some regimes, but in others may break down to produce "chaotic attractors," which can be demonstrated to have fractal properties (Kaplan and Glass, 1995). We discuss this further in the next section.

- Fractal objects may also be produced by large, complex systems. Based largely on computer experiments with "cellula automata," (models with large numbers of interacting "cells," whose behavior follows simple rules), it has been argued that such systems tend "to evolve into a poised, "critical" state, way out of balance, where minor disturbances may lead to events...of all sizes...The state is established solely because of the dynamical interactions among individual elements of the system: the critical state is *self-organized*." (Bak, 1996, p.1–2). The phenomena produced show power-law scaling, and fractal or multifractal characteristics. Bak has claimed that such models explain the characteristics of such diverse phenomena as earthquakes, avalanches, forest fires, mass extinctions, and punctuated equilibria. He has argued that self-organized criticality, developed in systems with large numbers of degrees of freedom, is the most probable explanation of most natural power-law phenomena (which, he claims, cannot be produced by low-dimensional chaotic systems).

- Complex fingering patterns with fractal geometry may be produced when a fluid spreads through a porous medium. This process (and its physical analogs) is known as *percolation* (Feder, 1988, Chapter 7). Somewhat similar patterns are produced by stochastic growth processes, such as *diffusion-limited aggregation*: some of these patterns are very similar to natural dendrites and skeletal crystals (see Meakin and Fowler, in Barton and Lapointe, 1995a, p.227–261). There is now a large body of theory, experiment, and numerical simulation that explains how these patterns form.

It is clear that, at present, the construction of theoretical models that explain how fractals form is much better developed than techniques for determining the existence or properties of natural fractals. This does not seem inappropriate: for scientists, understanding how fractal objects form is more interesting and important than just being able to describe them. For example, understanding how sand is formed, transported, and deposited is more important than describing the detailed size distribution of sand deposits.

10.5   Nonlinear Dynamics and Chaos

Dynamics is used here to refer to any system whose behavior is controlled by a system of difference, or differential equations. Dynamic systems include pendulums and convecting fluids, electrical circuits, lasers, chemical systems, and models of biological systems. Most of the systems of interest are nonlinear, that is the governing equations include nonlinear terms (such as powers and cross-products of the variables), and can be described by a relatively small set of first order, ordinary differential (or difference) equations.

Chaos has become a popular, though poorly defined word, to describe the apparently random behavior of some nonlinear dynamic systems. Very simple sets of equations may give rise to very complex behavior. In theory, this behavior is completely determined by the starting conditions, but in practice it is not, because chaotic systems show extreme sensitivity to starting conditions (and computational approximations). So chaotic systems can only be predicted a short time into the future. In this section we describe two classic examples of chaotic systems.

The first example is the chaotic behavior of the **Logistic difference equation**. This equation, also known as the *logistic map* is

$$x_{i+1} = rx_i(1 - x_i) \tag{10.7}$$

It can be used as a model for the growth of biological populations: $x = 1$ is the limiting size of the population, and the equation shows how the population grows or contracts from one generation to the next. r is a parameter in the range $0 < r \leq 4$. The equation is extensively discussed in most texts on nonlinear dynamics. The following is a MATLAB function that displays the values of x returned by the logistic equation, for a specified range of r values. The starting value is arbitrarily set at 0.3, and the first 200 values of x are rejected as "transients," followed by points that show a "steady state." The characteristics of the "steady state" depend on the value of r, but generally not on the starting value of x—they are therefore described as an "attractor" of the system.

```
function x = logist(r1, r2)
% x = logist(r1,r2)
% calculates the logistic map
%     x(i+1) = r*x(i)*(1-x(i))
% for 100 values of x from r1 to r2,
% increasing by steps of dr = r2-r1/200
% The first 200 values of x for each r
% are discarded, and the next 100 plotted.
% Written by Gerry Middleton, Jan 1997
x = zeros(1:300,1);
dr = (r2-r1)/200;
rj = r1;
plot(r1,0.3,'o'), title('Logistic Map');
axis([r1 r2 0 1]);
xlabel('r'), ylabel('x');
hold on;
for j = 1:201
    x(1) = 0.3;
    for i = 2:300
        x(i) = rj*x(i-1)*(1-x(i-1));
    end;
    r = rj*(ones(100,1));
    plot(r,x(201:300),'.y');
    rj = rj+dr;
end
x = x(201:300);
hold off
```

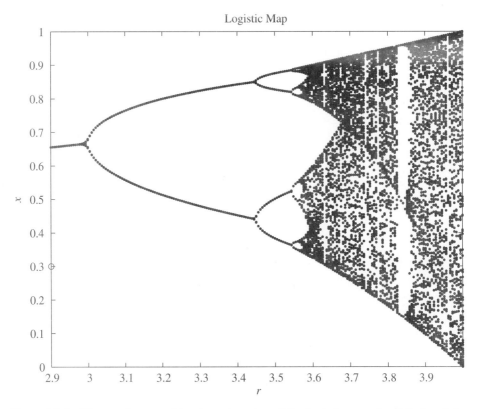

Figure 10.5   The x values resulting from iteration of the logistic map, for $2.9 \leq r \leq 4$.

An example of the output is shown in Figure 10.5. Note that MATLAB is not the ideal computing environment for this type of system, because the value of x_{i+1} cannot be computed until x_i is known: there is no way that loops can be "vectorized" for such a system, so MATLAB loses much of its power. The vector x returned by the function includes the set of x values for the set upper limit of r. For $r = 4$ it has been shown that these have the property of a randomized version of the well-known fractal called "Cantor's dust." Thus, the attractor set produced by the logistic map has fractal properties for some values of r—also, some parts of the map show self-similarity in that they look very much like the whole map when enlarged.

The second example of a dynamical system exhibiting chaos is the **Lorenz system**. This system consists of a set of three nonlinear differential equations:

$$\frac{dx}{dt} = s(y - x) \tag{10.8}$$

$$\frac{dy}{dt} = rx - y - xz \tag{10.9}$$

$$\frac{dz}{dt} = xy - bz \tag{10.10}$$

The system was devised by Lorenz in 1963 as the simplest possible model for thermal convection in the atmosphere. It is discussed in detail in many books on nonlinear dynamics (e.g., see Middleton et al., 1995, p.34–38, and references cited in that book). s and b are generally treated as constants with the values 10 and $8/3$ respectively. r can be considered to be a controlling parameter (the ratio of the Rayleigh number to the critical Rayleigh number at which convective motion begins). Chaotic regimes are shown for several different ranges of r.

To implement the Lorenz system in MATLAB, or to implement other systems defined by ordinary differential equations, we make use of MATLAB's built-in functions, `ode23` and `ode45`, for solving such sets of equations. These functions implement different forms of the Runge-Kutta numerical method for solving differential equation. The details of this method need not concern us (but see Lindfield and Penny, 1995, Chapter 5): to use the MATLAB functions one must first construct a function defining the set of equations (in this case, `flrnz`) and then call it in the way described in the MATLAB manual. The following two functions show how this is done.

```
function fu = flrnz(t,x)
% fu = flrnz(t,x)
% defines the Lorenz equations
% t is time and x is a vector with
% three components, fu is the vector
% of their derivatives wrt t
% Written by Gerry Middleton, Jan 1997
global r;    % a control coefficient
fu = zeros(3,1);
sig = 10;    % a second coefficient
b = 8/3      % and a third
fu(1) = sig*(x(2) - x(1));
fu(2) = r*x(1) - x(2) - x(1)*x(3);
fu(3) = x(1)*x(2) - b*x(3);


function X = lorenz2(r, tf, x0, acc)
% X = lorenz2(r, tf, x0, acc)
% solves and displays the trajectory
% of the Lorenz equations, which are
% defined by function flrnz.m
% r is the ratio of the Rayleigh
% number to the critical value
% (default is 28); tf is the total time
% (50), x0 is the starting position
% acc is the accuracy (0.001)
% X contains the trajectory coordinates
```

```
% Written by Gerry Middleton, 1997 for MATLAB 4
% for v.5, modify the call to ode23 as commented
global r;
if nargin < 1
   r = 28;
end;
if nargin < 2
   tf = 50;
   x0 = [1 5 10];
end
fprintf('Please wait -- computing\n');
[t X] = ode23('flrnz', [0 tf], x0); % MATLAB 5 version
%[t X] = ode23('flrnz', 0, tf, x0); MATLAB 4 version
plot3(X(:,1),X(:,2), X(:,3),x0(1),x0(2),x0(3),'or');
view(20,30);
s = num2str(r);
title(['Lorenz trajectories for r = ' s]);
xlabel('x'), ylabel('y'), zlabel('z');
figure;
[N, c] = size(X);
if N > 1000
  N = 1000;
end
subplot(3,1,1);
plot(t(1:N),X(1:N,1));
ylabel('x');
subplot(3,1,2);
plot(t(1:N), X(1:N,2));
ylabel('y');
subplot(3,1,3);
plot(t(1:N), X(1:N,3));
ylabel('z'), xlabel('t');
```

Note that the value of r is entered as an argument to the main function: in order that this can be used by the function defining the equations, it must be declared as a "global" variable by both functions. The trajectory in xyz-space, that is the path travelled in space as time is incremented, is shown in Figure 10.6. Time series for the x, y, and z variables are shown in Figure 10.7. Note that x, y, and z are *not* spatial coordinates, so the trajectories in xyz-space do *not* correspond to paths of particles in the convecting fluid model. For an illustration of these paths see Middleton et al. (1995, p.37). The xyz-space is described as the *state space* of the system.

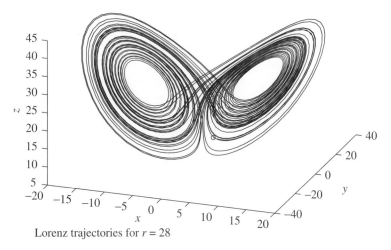

Lorenz trajectories for $r = 28$

Figure 10.6 Trajectories in the Lorenz system, for $r = 28$. The initial point is shown by the circle.

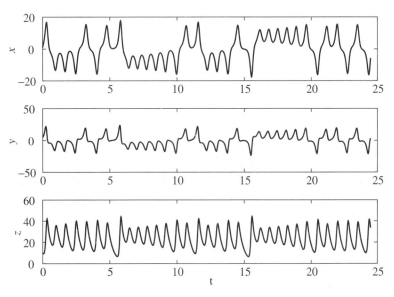

Figure 10.7 Time series for the first 1000 points in the Lorenz trajectories shown in Figure 10.6. Although these series may appear to show a regular periodicity, Fourier analysis of much longer time series produced by the Lorenz attractor shows that no such periodicity exists.

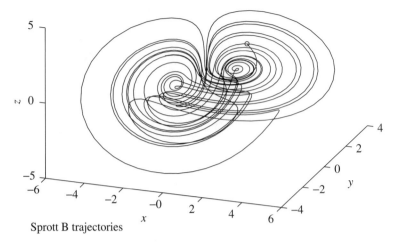

Sprott B trajectories

Figure 10.8   Trajectories in the Sprott B system. The initial point is shown by the circle. Note the increase in complexity, reflected by the higher fractal dimension (2.17) in the Sprott B system.

It can be seen from Figure 10.7 that the time series produced by the Lorenz equations are smooth at small scales, i.e., they are *not* fractals. Nevertheless, the attractor itself (defined in theory by an infinite number of trajectories on its surface) can be shown to be fractal, with a dimension of 2.06. Most of the well known fractals produced by differential equations and illustrated in texts on nonlinear dynamics, have fractal dimensions that are only a little larger than 2.0. In fact it is hard to find examples of more obviously fractal attractors in three dimensions. Sprott (1993, 1994) has described and illustrated a number that have fractal dimensions between 2.1 and 2.2. One (Sprott's B model) is shown in Figure 10.8: the corresponding time series are shown in Figure 10.9. The functions that generated these figures (`fsprb.m` and `sprottb.m`) are included with those distributed with this text, but not reprinted here, because they are very similar to those that generated the figures for the Lorenz system.

MATLAB includes a demo script `lorenz` that gives a more visually attractive demonstration of the Lorenz attractor. Careful study of the script which runs this demonstration will introduce the reader to some visual programming tricks that go beyond anything attempted in this text.

10.6   Detecting Chaos

For natural scientists, the existence of chaotic systems raises the question of how such systems may be identified, if only one or more measured time series generated by the system is available. Or to phrase it differently, how is a signal produced by a chaotic system different from a signal produced by a stochastic system?

The answer is provided by a remarkable theorem known as the *embedding theorem*.

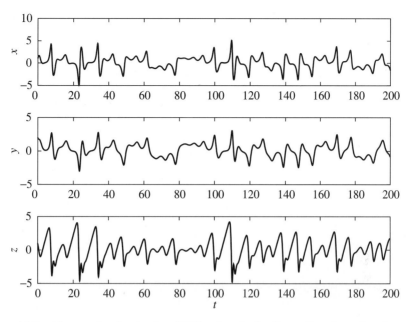

Figure 10.9   Time series for the first 2000 points in the Sprott B trajectories shown in Figure 10.8. Note the similarities to and differences from the Lorenz examples.

The process of *embedding* converts a one-dimensional time series into a d-dimensional time series, generally by using the *method of delays*. Suppose the observed time series is $x(t)$, consisting of N observations

$$x(1), x(2), x(3), \ldots, x(N)$$

Then to embed the time series in d dimensions, we choose a lag equal to an integral number of time delays, m, and form the d time series shown in the columns of Table 10.1. This may be written more compactly as the series of coordinates of the embedding space

$$x(t), x(t+T), x(t+2T), \ldots, x(t+(d-1)T).$$

The data is changed from a single vector with one column, and N rows, to a matrix with d columns and $n = (N - (d-1)m)$ rows. In each column, t increments by Δt. From one column to the next t increments by $T = m\Delta t$.

The Takens embedding theorem states that, under rather general conditions, the topological properties of the original attractor can be reconstructed from an embedding in d_e space, provided that $d_e \geq 2d_a$ where d_a is the topological dimension of the attractor (in general, $k - 1 < d_a < k$, where k is the number of variables $x_1(t), x_2(t), \ldots, x_k(t)$ defining the state space of the dynamical system). The theory of embedding makes use of fairly advanced concepts in topology (for a brief explanation and reprints of original papers see Ott et al., 1994).

$$x(1), \quad x(1+m), \quad x(1+2m), \quad \dots, \quad x(1+(d-1)m)$$
$$x(2), \quad x(2+m), \quad x(2+2m), \quad \dots, \quad x(2+(d-1)m)$$
$$x(3), \quad x(3+m), \quad x(3+2m), \quad \dots, \quad x(3+(d-1)m)$$
$$\dots \qquad \dots \qquad \dots \qquad \dots \qquad \dots$$
$$x(n), \quad x(n+m), \quad x(n+2m), \quad \dots, \quad x(n+(d-1)m)$$

Table 10.1   Time series produced by embedding in d-dimensional space using a lag of m. Each column gives the coordinates in a different dimension, for a total of n points.

A suitable choice of lag time can produce trajectories from a single time series which bear a striking resemblance to the trajectories produced by the entire set of equations. This is shown for the Lorenz system, in Figure 10.10. The input time series consisted of the x-values (only) plotted in Figures 10.6 and 10.7 (and saved in lrnz50.dat). The appropriate choice of lag time is discussed by Ott et al. (1994) and Abarbanel (1996). As a first guess, one may use the first minimum in the autocorrelation function.

The following MATLAB script embeds a time series, and displays the trajectories, if $d \leq 3$.

```
function X = embed(x, lag, dim)
% X = embed(x, lag, dim)
% inputs x, a vector of data containing
% N values, in a space of dimension dim
% (default is 3), using a delay of lag
% and creates a matrix X whose rows are:
% [x(1) x(1+lag) x(1+2*lag)...x(1+dim*lag)]
% [x(2) x(2+lag) x(2+2*lag)...x(2+dim*lag)]
% ....................................
% [x(n) x(n+lag) x(n+2*lag)...x(n+dim*lag)]
% where n = N - (dim-1)*lag. If dim=3, it
% plots views from different azimuths.
% Written by Gerry Middleton, Jan. 1997
if nargin < 3
   dim = 3;
end
N = length(x);
n = N - (dim-1)*lag;
for i = 1:dim
   X(1:n,i) = [x( (1+(i-1)*lag):(n + (i-1)*lag) )];
end;
if dim == 3
   for az = -75:30:75
      plot3(X(:,1), X(:,2),X(:,3));
```

Embedded time series

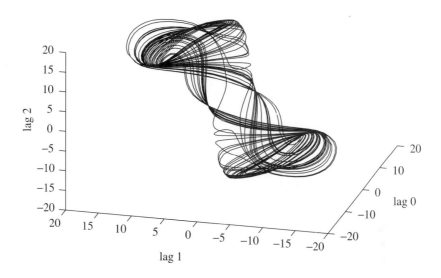

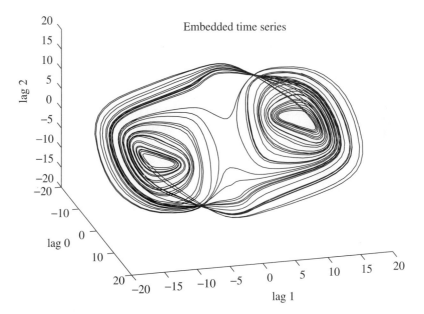

Figure 10.10   Two views of the Lorenz attractor, reconstructed from a time series, using the method of delays.

```
        title('Embedded time series');
        xlabel('lag 0'),ylabel('lag 1'),zlabel('lag 2');
        view(az,30);
        if az < 75
            figure;
        end;
    end;
end
```

If the system has an attractor with a dimension less than 3, its existence is generally apparent from the plot of a time series in three embedding dimensions, as shown by the two views in Figure 10.10. The function plots views from more than one direction: it is not obvious that the basic topology of the Lorenz attractor is preserved in the view shown in the upper part of the figure, but the view shown in the lower part of the figure reveals the similarity more clearly. If no attractor was present (or if it had a dimension larger than 3) then the trajectories would appear as a confused mass of intersecting lines.

For embedding dimensions larger than three it is no longer possible to view the trajectories directly. Various numerical techniques have been developed to reveal the presence of an attractor, if one exists. The techniques go beyond the scope of this book, but they are described in Abarbanel (1996), Ott et al. (1994), and Middleton et al. (1995).

10.7   Software

There is an abundance of software for those who wish to explore fractals and nonlinear dynamics. A popular freeware program, constantly adding new features, `fractint`, available from many sites. A site specializing in fractal software and images is `spanky.triumph.ca`. The Usenet groups `sci.fractals` and `sci.nonlinear` produce good FAQs (frequently asked questions) that include up-to-date lists of software. The author has prepared several MATLAB scripts for fractal analysis not described in this chapter. They are available from The MathWorks, Inc. ftp site `ftp.mathworks.com/pub/contrib/math/fractals`. Scheinerman (1996) is a text that includes numerous MATLAB programs for both fractals and nonlinear dynamics. Middleton et al. (1995) include a diskette of nonlinear dynamics programs for the PC, and a bibliography of other software.

10.8   Recommended Reading

Bak, Per, 1996, How Nature Works: The Science of Self-organized Criticality. Springer-Verlag, 212 p. (QC173.4.C74B34. Applications to avalanches, earthquakes, topography, evolution, and extinction: try not to be put off by the gratuitous insults heaped by the author on earth scientists!)

Brown, S.R., 1995, Measuring the dimension of self-affine fractals: example of rough surface: in Barton, C.C., and P.R. LaPointe, eds., 1995, Fractals in the Earth Sciences. New York, Plenum, p.77–87 (QE33.2.F73F73)

Falconer, K., 1990, Fractal Geometry: Mathematical Foundations and Applications. New York, Wiley, 288 p. (QA614.86.F35 A valuable book for the mathematically stout-at-heart.)

Feder, J., 1988, Fractals. New York, Plenum, 283 p. (QA447.J46 Still the best book on scientific applications.)

Hastings, H. M. and G. Sugihara, 1993, Fractals, A User's Guide for the Natural Sciences. Oxford University Press, 235 p. (QA614.86.H37 Mathematically fairly advanced, but includes useful discussions and implementations (in Pascal) of numerical demonstrations and techniques.)

Kaplan, D. and L. Glass, 1995, Understanding Nonlinear Dynamics. New York, Springer-Verlag, 420 p. (QA845.K36 A fine book, covering fractals and cellular automata as well as nonlinear dynamics at a mathematically elementary but sophisticated level.)

Korvin, G., 1992, Fractal Models in the Earth Sciences: Amsterdam, Elsevier, 396 p.(QE33.2.F73K67)

Mandelbrot, B., 1967, How long is the coast of Britain? Statistical self-similarity and fractional dimension. Science, v.156, p.636–638 (A classic paper).

Mandelbrot, B., 1982, The Fractal Geometry of Nature: New York, W.H. Freeman and Co, 468 p. (QA447.M357)

Middleton, G.V., R.E. Plotnick and D.M. Rubin, 1995, Nonlinear Dynamics and Fractals: New Numerical Techniques for Sedimentary Data. SEPM Short Course No. 36, 174 p. (QE471.M54)

Peitgen, H-O., H. Jürgens and D. Saupe, 1992, Chaos and Fractals: New Frontiers of Science: New York, Springer-Verlag, 984 p. (QA614.86.P43 Well-written text with excellent illustrations; little math background required.)

Scheinerman, E.R., 1996, Invitation to Dynamical Systems. Upper Saddle River NJ, Prentice Hall, 373 p. (QA614.8.S34 A text on linear and nonlinear dynamical systems and fractals, including many MATLAB scripts.)

Schroeder, M., 1991, Fractals, Chaos, Power Laws. New York, W.H. Freeman, 429 p. (QC174.17.S9S38)

Sprott, J.C., 1993, Strange Attractors: Creating Patterns in Chaos. M&T, 426 p.+ diskette. (QA614.86.s67 A picture book of attractors, with programs for making them.)

Turcotte, D.L., 1992, Fractals and Chaos in Geology and Geophysics. Cambridge University Press, 221 p. (QE33.2.M3T87)

Turcotte, D.L., 1994, Fractal aspects of geomorphic and stratigraphic processes. GSA Today, v. 4, p. 201, 211–213.

References

Abarbanel, H.D.I., 1996, Analysis of Observed Chaotic Data. New York, Springer, 272 p. (Q172.5.C45A23)

Afifi, A.A., and V. Clark, 1990, Computer-Aided Multivariate Analysis. New York, Van Nostrand Reinhold, second edition, 505 p. (QA278.A33)

Aitchison, J., 1984, Statistical analysis of geochemical compositions. Mathematical Geology, v.16, p.531–564.

Aitchison, J., 1986, The Statistical Analysis of Compositional Data. London, Chapman and Hall, 416 p. (QA278.A37)

Albarède, F., 1995, Introduction to Geochemical Modeling. Cambridge University Press, 543 p. (QE515.A53)

Anderson, T.W., 1984, An Introduction to Multivariate Statistical Analysis. New York, John Wiley and Sons, second edition, 675 p. (QA278.A516)

Andrle, Robert, 1992, Estimating fractal dimension with the divider method in geomorphology. Geomorphology, v.5, p.131–141.

Andrle, Robert, 1996, Complexity and scale in geomorphology: Statistical self-similarity vs. characteristic scales. Mathematical Geology, v.28, p.275–293.

Bak, Per, 1996, How Nature Works: The Science of Self-organized Criticality. Springer-Verlag, 212 p. (QC173.4.C74B34)

Barnard, M.M., 1935, The secular variation of skull characteristics in four series of Egyptian skulls. Annals of Eugenics, v.6, p.352–371.

Barndorff-Nielsen, O.E. and C. Christiansen, 1988, Erosion, deposition and size distribution of sand. Proceedings Royal Society of London, v.A417, p.335–352.

Barth, T.F.W., 1948, Oxygen in rocks: a basis for petrographic calculations. Journal of Geology, v. 56, p.50–60.

Barton, C. and P.R. Lapointe, eds., 1995a, Fractals in the Earth Sciences: New York, Plenum, 265 p. (QE33.2.F73F73)

Barton, C. and P.R. Lapointe, eds., 1995b, Fractals in Petroleum Geology and Earth Processes: New York, Plenum, 317 p.(TN870.53.F73)

Becker, R., J. Chambers and A. Wilks, 1988, The New S Language. Belmont CA, Wadsworth. (QA76.73.S15B43)

Bendat, J.S. and A.G. Piersol, 1971. Random Data: Analysis and Measurement Procedures. N.Y., Wiley-Interscience, 407 p. (TA340.B43)

Bennett, R.J., 1979, Spatial Time Series: Analysis—Forecasting—Control. London, Pion, 674 p. (QA280.B46)

Berry, D.A., 1996, Statistics: A Bayesian Perspective. Belmont, CA, Duxbury, 519 p. (QA279.5.B48)

Berry, D.A. and B.W. Lindgren, 1996, Statistics: Theory and Methods. Belmont CA, Duxbury, second edition, 702 p. (QA276.12.B48)

Bevington, P.R. and D.K. Robinson, 1992, Data Reduction and Error Analysis for the Physical Sciences. New York, McGraw-Hill, second edition, 328 p. (QA278.B48)

Blatt, H., G. Middleton and R. Murray, 1980, Origin of Sedimentary Rocks. Englewood Cliffs NJ, Prentice-Hall, second edition, 782 p. (QE471.B65)

Box, G.E.P. and G.N. Jenkins, 1994, Time Series Analysis: Forecasting and Control. Englewood Cliffs NJ, Prentice-Hall, third edition, 598 p. (QA280.B67)

Bras, R.L. and I. Rodríguez-Iturbe, 1985, Random Functions and Hydrology. New York, Dover, 559 p. (GB656.2.M33B73)

Brockwell, P.J. and R.A. Davis, 1991, Time Series: Theory and Methods. New York, Springer-Verlag, second edition, 577 p. (QA280.B76)

Brooker, Peter I., 1991, A Geostatistical Primer. Teaneck NJ, World Scientific, 95 p. (QE33.2.S82B75)

Brown, S.R., 1995, Measuring the dimension of self-affine fractals: example of rough surface: in Barton, C.C., and P.R. LaPointe, eds., 1995, Fractals in the Earth Sciences. New York, Plenum, p.77–87 (QE33.2.F73F73)

Burroughs, W.J., 1992, Weather Cycles: Real or Imaginary? Cambridge University Press, 207 p. (QC883.2.C5B87)

Caceci, M.S. and W.P. Cacheris, 1984, Fitting curves to data: The simplex algorithm is the answer. Byte, May 1984, p.340–362.

Carr, J.R., 1995, Numerical Analysis for the Geological Sciences. Englewood Cliffs NJ, Prentice-Hall, 592 p. (QE33.2.M3C37)

Chatfield, C., 1989. The Analysis of Time Series: an Introduction. London, Chapman and Hall, fourth edition, 241 p. (QA280.C4)

Chayes, Felix, 1956, Petrographic Modal Analysis: An Elementary Statistical Appraisal. New York, John Wiley and Sons, 113 p. (QE434.C51)

Chayes, Felix, 1971, Ratio Correlation. University of Chicago Press, 99 p. (QE431.5.C47)

Cheeney, R.F., 1983, Statistical Methods in Geology. London, George Allen and Unwin, 169 p. (QE33.2.M3C48)

Clayton, R.N., I. Friedman, D.L. Graf, T.K. Mayeda, W.F. Meents, and N.F. Shimp, 1966, The origin of saline formation waters, I. Isotopic composition. Journal of Geophysical Research, v.71, p.3869–3882.

Cochran, W.G., F. Mosteller and J.W. Tukey, 1954, Appendix G, Principles of sampling, *in* Statistical Problems of the Kinsey Report. American Statistical Association, p.309–331.

Cox, Allan and R.B. Hart, 1986, Plate Tectonics: How It Works. Palo Alto, Blackwell Sci., 392 p. (QE511.4.C683)

Cressie, N.A.C., 1993, Statistics for Spatial Data. N.Y., John Wiley and Sons, revised edition, 900 p. (QA278.2.C75)

Cutler, C.D., 1994, A review of the theory and estimation of fractal dimension. in H. Tong, ed., Nonlinear Time Series and Chaos, v.1: Dimension Estimates and Models. Singapore, World Scientific, p.1–107.

Davis, J.C., 1986, Statistics and Data Analysis in Geology. New York, John Wiley and Sons, second edition, 646 p. (QE48.8.D38)

Deutsch, C.V. and A.G. Journel, 1992, GSLIB: Geostatistical Software Library and User's Guide. Oxford University Press, 340 p.+ diskette. (QE33.2.S82D48)

De Wijs, H.J., 1951, Statistics of ore distribution. Part I. Frequency distribution of assay values. Geologie en Mijnbouw, v.13, no.11, p.365–375.

Dickinson, W.R. and C.A. Suczek, 1979, Plate tectonics and sandstone compositions. American Association of Petroleum Geologists Bulletin, v.63, p.2164–2183.

Draper, N.R. and H. Smith, 1981, Applied Regression Analysis. New York, John Wiley and Sons, second edition, 709 p. (QA278.2.D7)

Dunn, G. and B.S. Everitt, 1982, An Introduction to Mathematical Taxonomy. Cambridge University Press, 152 p. (QH83.D86)

Etter, D.M., 1993, Engineering Problem Solving with MATLAB. Engelwood Cliffs, Prentice Hall, 434 p. (TA331.E88)

Eubank, Stephen and Doyne Farmer, 1990, An introduction to chaos and randomness *in* Jen, Erica, ed., 1989 Lectures in Complex Systems. Reading, MA, Addison-Wesley Publ. Co., Inc., Santa Fe Institute Studies in the Sciences of Complexity, Lectures v.II, p.75–190 (QA267.7.L44).

Evans, M. and N. Hastings, 1993, Statistical Distributions. New York, Wiley Interscience, second edition, 170 p. (QA273.6.E92)

Fairbairn, H.W., 1951, A cooperative investigation of precision and accuracy in chemical, spectrochemical and modal analysis of silicate rocks. U.S. Geological Survey Bulletin 980.

Falconer, K., 1990, Fractal Geometry: Mathematical Foundations and Applications. New York, Wiley, 288 p. (QA614.86.F35)

Feder, J., 1988, Fractals. New York, Plenum, 283 p. (QA447.J46)

Fedikow, M.A.F., D. Parberry and K.J. Ferreira, 1991, Geochemical target selection along the Agassiz metallotect utilizing stepwise discriminant function analysis. Economic Geology, v.86, p.588–599.

Ferguson, J., 1994, Introduction to Linear Algebra in Geology. London, Chapman and Hall, 203 p. (QE33.2.M3F47)

Fisher, N.I., 1993, Statistical Analysis of Circular Data. Cambridge University Press, 277 p. (QA276.F488)

Fisher, N.I., T. Lewis and B.J.J. Embleton, 1987, Statistical Analysis of Spherical Data. Cambridge University. Press, 329 p. (QA276.F489 1987)

Fisher, R.A., 1936, The use of multiple measurements in taxonomic problems. Annals of Eugenics, v.7, p.179–188. (reprinted in Collected Papers of R.A. Fisher, v.3, p.466–475).

Forsyth, D. and S. Uyeda, 1975, On the relative importance of the driving forces of plate tectonics. Geophysical Journal of the Royal Astronomical Society, v.43, p.163–200.

Garcia, A.L., 1994, Numerical Methods for Physics. Englewood Cliffs NJ, Prentice Hall, 368 p. (QC20.G37)

Gellert, W., H. Küstner, M. Hellwich and H. Kästner, eds., 1975, The VNR Concise Encyclopedia of Mathematics. New York, Van Nostrand Reinhold, 760 p. (second edition, 1989. QA40.V18)

Godin, G., 1972, The Analysis of Tides. University of Toronto Press, 264 p. (QB415.2.G6)

Goldich, S.S., 1938, A study in rock weathering. Journal of Geology, v.46, p.17–58.

Gottman, J.M., 1981, Time Series Analysis: a Comprehensive Introduction for Social Scientists. Cambridge University Press, 400 p. (HA30.3 G67)

Griffiths, J.C., 1967, Scientific Method in the Analysis of Sediments. New York, McGraw-Hill, 508 p. (QE433.G7)

Haan, C.T., 1977, Statistical Methods in Hydrology. Ames IA, Iowa State University Press, 378 p. (GB656.2.S7H3)

Hanselman, D. and B. Littlefield, 1996, Mastering MATLAB. Englewood Cliffs NJ, Prentice-Hall, 542 p. (QA297.H298)

Hanselman, D. and B. Littlefield, 1998, Mastering MATLAB 5. Englewood Cliffs NJ, Prentice-Hall, 638 p. (QA297.H296)

Harbaugh, J.W. and D.F. Merriam, 1968, Computer Applications in Stratigraphic Analysis. New York, John Wiley and Sons, 282 p. (QE652.H3)

Hastings, H. M. and G. Sugihara, 1993, Fractals, A User's Guide for the Natural Sciences. Oxford University Press, 235 p. (QA614.86.H37)

Hayes, Monson B., 1996, Statistical Digital Signal Processing and Modeling. New York, John Wiley and Sons, 608 p.

Hegge, B.J. and G. Masselink, 1996, Spectral analysis of geomorphic time series: auto-spectrum. Earth Surface Processes and Landforms, v.21, p.1021–1040.

Hobbs, B.E., W.D. Means and P.F. Williams, 1976. An Outline of Structural Geology. New York, John Wiley and Sons, 571 p. (QE601.H6)

Hohn, M.E., 1988, Geostatistics and Petroleum Geology. New York, Van Nostrand Reinhold, 264 p. (TN871.H536)

Howard, P.J.A., 1991, An Introduction to Environmental Pattern Analysis. Park Ridge NJ, Parthenon, 254 p. (TD193.H69)

Hubbard, B.B., 1996, The World According to Wavelets: The Story of a Mathematical Technique in the Making. Wellesley MA, A.K. Peters, 264 p. (QA403.3H83)

Imbrie, J., 1963, Factor and Vector Analysis Programs for Analyzing Geologic Data. Evanston IL, Northwestern University, Tech. Rept. 6, ONR Task No. 389-135, 83 p.

Isaaks, E.H. and R.M. Srivastava, 1989, An Introduction to Applied Geostatistics. Oxford University Press, 561 p. (QE33.2.M3I83)

Jones, T.A. and D.E. Hamilton, 1986, Contouring Geologic Surfaces with the Computer. New York, Van Nostrand Reinhold, 314 p. (QE36.J66)

Journel, A.G., 1986, Geostatistics: Models and tools for the earth sciences. Mathematical Geology, v.18, p.119–140.

Journel, A.G. and C.J. Huijbregts, 1978, Mining Geostatistics. London, Academic, 600 p. (TN260.J68)

Kaplan, D. and L. Glass, 1995, Understanding Nonlinear Dynamics. New York, Springer-Verlag, 420 p. (QA845.K36)

Kay, Steven M., 1988, Modern Spectral Estimation: Theory and Application. Englewood Cliffs, NJ, Prentice-Hall, 543 p. (QA280.K39)

Kendall, M.G., 1959, Hiawatha designs an experiment. American Statistician, v.13, no.5, p.23–24.

Kent, J.T., G.S. Watson and T.C. Onstott, 1990, Fitting straight lines and planes with an application to radiometric dating. Earth and Planetary Science Letters, v.97, p.1–17.

Kinsman, Blair, 1965, Wind Waves: their Generation and Propagation on the Ocean Surface. Englewood Ciffs, NJ, Prentice-Hall Inc., 676 p. (GC211.K5)

Kitanidis, P.K., 1997, Introduction to Geostatistics: Applications in Hydrogeology. Cambridge University Press, 259 p. (GB1001.72.57K57)

Klinkenberg, Brian, 1994, A review of methods used to determine the fractal dimension of linear features. Mathematical Geology, v.26, p.23–46.

Kolmogoroff, A.N., 1963, The theory of probability. Chapter 11, p.229–264 *in* Aleksandrov, A.D., A.N. Kolmogoroff and M.A. Lavrent'ev, eds., Mathematics: Its Content, Methods, and Meaning, v.2, reprinted 1990 by Dorset Press, New York. (QA36.A454)

Korvin, G., 1992, Fractal Models in the Earth Sciences: Amsterdam, Elsevier, 396 p. (QE33.2.F73K67)

Krumbein, W.C., 1942, Flood deposits of Arroyo Seco, Los Angeles County, California. Bulletin of the Geological Society of America, v.53, p.1355-1402.

Krumbein, W.C. and F.A. Graybill, 1965, An Introduction to Statistical Models in Geology. New York, McGraw-Hill, 475 p. (QE40.K94)

Lancaster, P. and K. Šalkauskas, 1986, Curve and Surface Fitting: An Introduction. New York, Academic Press, 280 p. (QA297.6)

Le Maitre, R.W., 1982, Numerical Petrology: Statistical Interpretation of Geochemical Data. New York, Elsevier, (QE431.5.L33)

Lindfield, G. and J. Penny, 1995, Numerical Methods Using MATLAB. New York, Ellis Horwood, 328 p. (QA297.P45)

Lowry, Richard, 1989, The Architecture of Chance: an Introduction to the Logic and Arithmetic of Probability. Oxford University Press, 175 p. (QA273.L685)

Mandelbrot, B., 1967, How long is the coast of Britain? Statistical self-similarity and fractional dimension. Science, v.156, p.636–638.

Mandelbrot, B., 1982, The Fractal Geometry of Nature. New York, W.H. Freeman and Co, 468 p. (QA447.M357)

Mandelbrot, B., 1989, Les Objets Fractals: Forme, Hasard et Dimension. Paris, Flammarion, third edition, 268 p.

Marcotte, Denis, 1991, Cokriging with MATLAB. Computers and Geosciences, v. 17, p.1265–1280.

Marcotte, Denis, 1996, Fast variogram computation with FFT. Computers and Geosciences, v.22, p.1175–1186.

Marple, S.L., Jr., 1987, Digital Spectral Analysis with Applications. Englewood Cliffs, NJ, Prentice-Hall, 492 p. (QA280.H38)

McBratney, A.B. and R. Webster, 1986, Choosing functions for semi-variograms of soil properties and fitting them to sampling estimates. Journal of Soil Science, v.37, p.617–639.

Medak, F. and N. Cressie, 1991, Confidence regions in ternary diagrams based on the power-divergence statistics. Mathematical Geology, v.23, p.1045–1055.

Middleton, G.V., 1964, Statistical studies on scapolites. Canadian Journal of the Earth Sciences, v.1, p.23–34.

Middleton, G.V., 1990, Fitting cumulative curves using splines. Journal of Sedimentary Petrology, v.60, p.615–616.

Middleton, G.V., R.E. Plotnick and D.M. Rubin, 1995, Nonlinear Dynamics and Fractals: New Numerical Techniques for Sedimentary Data. SEPM Short Course No. 36, 174 p. (QE471.M54)

Middleton, G.V. and J.B. Southard, 1984, Mechanics of Sediment Movement. Soc. Economic Paleontologists Mineralogists Short Course Notes 3, second edition, 401 p. (QE471.2.M53)

Middleton, G.V. and P.R. Wilcock, 1994, Mechanics in the Earth and Environmental Sciences. Cambridge University Press, 459 p. (QA808.M49)

Miller, R.L. and J.S. Kahn, 1962, Statistical Analysis in the Geological Sciences. John Wiley and Sons, 483 p. (QE40.M64)

Moiola, R.J., A.B. Spencer and D. Weiser, 1974, Differentiation of modern sand bodies by linear discriminant analysis. Transactions of the Gulf Coast Association of Geological Societies, v.24, p.321–326.

Molinaroli, E., M. Blom and A. Basu, 1991, Methods of provenance determination tested with discriminant function analysis. Journal of Sedimentary Petrology, v.61, p.900–908.

Moss, B.E., L.A. Haskin, R.F. Dymek and D.M. Shaw, 1995, Redetermination and reevaluation of compositional variations in metamorphosed sediments of the Littleton Formation, New Hampshire. American Journal of Science, v.295, p.988–1019.

Munk, W.H. and D.E. Cartwright, 1966. Tidal Spectroscopy and Prediction. Philosophical Transactions of the Royal Society of London, v. 299A, p.533-581.

Newland, D.E., 1993, An Introduction to Random Vibrations, Spectral and Wavelet Analysis. New York, John Wiley, third edition, 477 p. (QA935.N46)

Norman, G.R. and D.L. Streiner, 1986, PDQ Statistics. Toronto, B.C. Decker, 172 p. (QA276.12.N67)

Okabe, A., B. Boots and K. Sugihara, 1992, Spatial Tessellations: Concepts and Applications of Voronoi Diagrams. New York, John Wiley and Sons, 532 p. (QA278.2.O36)

Oppenheim, A.V. and A.S. Willsky, 1983, Signals and Systems. Englewood Cliffs NJ, Prentice-Hall, 796 p. (QA402.O63)

Ott, E., T. Sauer and J.A. Yorke, eds., 1994, Coping with Chaos: Analysis of Chaotic Data and the Exploitation of Chaotic Systems. New York, John Wiley and Sons, 418 p. (Q172.5.C45C67)

Panchen, A.J., 1992, Classification, Evolution, and the Nature of Biology. Cambridge University Press, 403 p. (QH83.P35)

Pankratz, A., 1983, Forecasting with Univariate Box-Jenkins Models. New York, John Wiley, 562 p. (QA280.P37)

Parks, J.M., 1966, Cluster analysis applied to multivariate geologic problems. Journal of Geology, v.74, p.703–715.

Pearce, T.H., 1968, A contribution to the theory of variation diagrams. Contributions to Mineralogy and Petrology, v.19, p.142–157.

Peitgen, H-O., H. Jürgens and D. Saupe, 1992, Chaos and Fractals: New Frontiers of Science: New York, Springer-Verlag, 984 p. (QA614.86.P43)

Percival, D.B., and A.T. Walden, 1993, Spectral Analysis for Physical Applications: Multitaper and Conventional Univariate Techniques. Cambridge University Press, 583 p. (QA320.P434)

Philip, G.M., C.G. Skilbeck and D.F. Watson, 1987, Algebraic dispersion fields on ternary diagrams. Mathematical Geology, v.19, p.171–181.

Philip, G.M. and D.F. Watson, 1986, Matheronian geostatistics—Quo vadis? Mathematical Geology, v.18, p.93–117.

Philip, G.M. and D.F. Watson, 1988a, Determining the representative composition of a set of sandstone samples. Geological Magazine, v.125, p.267–272.

Philip, G.M. and D.F. Watson, 1988b, Angles measure compositional differences. Geology, v.16, p.976–979.

Potter, P.E., N.F. Shimp and J. Witters, 1963, Trace elements in marine and freshwater argillaceous sediments. Geochimica et Cosmochimica Acta, v.27, p.669–694.

Press, W.H. et al., 1989, Numerical Recipes in Pascal: the Art of Scientific Computing. Cambridge University Press, 759 p. (QA76.73.P2N87)

Purdy, E.G., 1963, Recent calcium carbonate facies of the Great Bahama Bank: 1. Petrography and reaction groups. Journal of Geology, v.71, p.334–355.

Ragan, D.M., 1985. Structural Geology: An Introduction to Geometrical Techniques. New York, John Wiley and Sons, Third edition, 393 p. (QE601.R23)

Reyment, R.A., 1989, Compositional data analysis. Terra Nova, v.1, p.29–34.

Reyment, R.A. and K.G. Jöreskog, 1993, Applied Factor Analysis in the Natural Sciences. Cambridge University Press, 371 p. (QA278.5.R49)

Ripley, B.D. 1981, Spatial Statistics. New York, John Wiley and Sons, 252 p. (QA278.2.R56)

Robinson, E.A., 1981, Time Series Analysis and Applications. Goose Pond Press. (QA280.R64)

Rollinson, H.R., 1992, Another look at the constant sum problem in geochemistry. Mineralogical Magazine, v.47, p.267–280.

Rollinson, H.R., 1993, Using Geochemical Data: Evaluation, Presentation, Interpretation. Harlow, Essex (England), Longman, 352 p. (QE515.R75)

Roser, B.P. and R.J. Korsch, 1988, Provenance signatures of sandstone-mudstone suites determined using discriminant function analysis of major-element data. Chemical Geology, v.67, p.119–136.

Russ, J., 1994, Fractal Surfaces: New York, Plenum, 309 p. (QC173.4.S94R88)

Russell, J.K. and C.R. Stanley, eds., 1990, Theory and Application of Pearce Element Ratios to Geochemical Data Analysis. Geological Association of Canada Short Course v.8, 315 p. (QE515.T49)

Savini, J. and G.L. Bodhaine, 1971, Analysis of current-meter data at Columbia River gauging stations, Washington and Oregon. U.S. Geological Survey Water Supply Paper 1869-F, 59 p.

Scheinerman, E.R., 1996, Invitation to Dynamical Systems. Upper Saddle River NJ, Prentice Hall, 373 p. (QA614.8.S34)

Schneider, G.M., S.W. Weingart and D.M. Perlman, 1982, An Introduction to Programming and Problem Solving with Pascal. New York, John Wiley, second edition, 468 p. (QA76.73.P2S36)

Schroeder, M., 1991, Fractals, Chaos, Power Laws. New York, W.H. Freeman, 429 p. (QC174.17.S9S38)

Schwarzacher, W., 1975, Sedimentation Models and Quantitative Stratigraphy. Amsterdam, Elsevier Publ. Co., 382 p. (QE471.S38)

Schwarzacher, W., 1987, Principles of quantitative lithostratigraphy. The treatment of single sections. Chapter IV.1 in Gradstein et al., Quantitative Stratigraphy (QE651.Q27)

Shaw, D.M. and J.D. Bankier, 1954, Statistical methods applied to geochemistry. Geochimica et Cosmochimica Acta, v.6, p.111–123.

Shaw, D.M. and A.M. Kudo, 1965, A test of the discriminant function in the amphibolite problem. Mineralogical Magazine, v.34, p. 423–435.

Shore, John, 1985, The Sachertorte Algorithm, and Other Antidotes to Computer Anxiety. New York, Penguin, 269 p. (QA76.9.C64S56)

Silver, S.D., 1970, Statistical Inference. London, Chapman and Hall, 191 p. (QA276.S5)

Size, W.B., ed., 1987, Use and Abuse of Statistical Methods in the Earth Sciences. Oxford University Press, 169 p. (QE33.2.M3U84)

Spector, P., 1994, An Introduction to S and S-Plus. Belmont CA, Duxbury, 286 p. (QA76.73.S15S63)

Spencer, A.B. and P.S. Clabaugh, 1967. Computer programs for fabric diagrams. American Journal of Science, v.265, p.166–172.

Sprott, J.C., 1993, Strange Attractors: Creating Patterns in Chaos. M&T, 426 p.+ diskette. (QA614.86.s67)

Sprott, J.C., 1994, Some simple chaotic flows. Physical Review E, v.50, no.2, p.R647–R650.

Steiglitz, K., 1996, A Digital Signal Processing Primer: with Applications to Digital Audio and Computer Music. Menlo Park CA, Addison-Wesley, 314 p. (TK5102.9.S74)

Stevens, S.S., 1946, On the theory of scales of measurement. Science, v.103, p.677–680.

Stokes, S., C.S. Nelson and T.R. Healey, 1989, Textural procedures for environmental discrimination of Late Neogene coastal and sand deposits, Southwest Aukland, New Zealand. Sedimentary Geology, v.61, p.135–150.

Strang, G., 1980, Linear Algebra and Its Applications. Englewood Cliffs NJ, Prentice-Hall, second edition. (third edition, 1988. QA184.S8)

Suppe, John, 1985. Principles of Structural Geology. Englewood Cliffs NJ, Prentice-Hall, 537 p. (QE601.S94)

Swann, A.R.H. and M. Sandilands, 1995, Introduction to Geological Data Analysis. Oxford, Blackwell Science, 446 p. (QE33.2.S82S93)

Tearpock, D.J. and R.E. Bischke, 1991, Applied Subsurface Geological Mapping. Englewood Cliffs NJ, Prentice-Hall, 648 p. (TN870.5.T38)

Thompson, A.H., 1991, Fractals in rock physics. Annual Review of Earth and Planetary Sciences, v. 19, p. 237–262.

Turcotte, D.L., 1992, Fractals and Chaos in Geology and Geophysics. Cambridge University Press, 221 p. (QE33.2.M3T87)

Turcotte, D.L., 1994, Fractal aspects of geomorphic and stratigraphic processes: GSA Today, v. 4, p. 201, 211–213.

Van Everdingen, D.A., J.A.M. Van Gool and R.L.M. Vissers, 1992, Quickplot: a microcomputer-based program for the processing of orientation data. Computers and Geosciences, v.18, p.283–287.

Vasey, G.M. and G.E. Bowes, 1985, The use of cluster analysis in the study of some non-marine bivalvia from the Westphalian D of the Sydney Coalfield, Nova Scotia. Journal of the Geological Society of London, v.142, p.397–410.

Watson, D.F., 1992, Contouring: A Guide to the Analysis and Display of Spatial Data. Oxford, Pergamon Press, 321 p. (GA125.W38)

Watson, D.F. and G.M. Philip, 1989, Measures of variability for geologic data. Mathematical Geology, v.21, p.233–254.

Waythomas, C.F. and G.P. Williams, 1988, Sediment yield and spurious correlation—towards a better portrayal of the annual suspended-sediment load of rivers. Geomorphology, v.1, p.309–316.

Whitten, E.H.T., 1995, Open and closed compositional data in petrology. Mathematical Geology, v.27, p.789–806.

Wirth, Niklaus, 1975, Algorithms + Data Structures = Programs. Engelwood Cliffs NJ, Prentice-Hall, 366 p. (QA76.6.W56)

Woodcock, N.H., 1977, Specification of fabric shapes using an eigenvalue method. Bulletin of the Geological Society of America, v.88, p.1231–1236.

Woronow, Alex, 1994, Identifying mineral controlling the chemical evolution of igneous rocks: Beyond Pearce element-ratio diagrams. Geochemica et Cosmochimica Acta, v.58, p.5479–5487.

Woronov, A. and K.M. Love, 1990, Quantifying and testing differences among means of compositional data suites. Mathematical Geology, v.22, p.837–852.

Yarus, J.M. and R.L. Chambers, eds., 1994, Stochastic Modeling and Geostatistics: Principles, Methods, and Case Studies. American Association of Petroleum Geologists Computer Applications in Geology, No. 3, 379 p.

Appendix A

Introduction to MATLAB

A.1 Introduction

MATLAB is a commercial data processing package that is becoming increasingly popular for scientific applications. It has the advantage of being available in a cheap Student Version: the student version is essentially the complete professional package, but with limitations on the size of data files. The Student Version also contains some routines taken from the MATLAB Signal Processing Toolbox (which must be purchased separately in the professional version). The version for WINDOWS, first released in 1995, has the advantage that graphs may be printed at high resolution. Further refinements were introduced with version 5 (for WINDOWS 95). The Student Version is also available in a version for Macintosh. The professional version is available for many other computers: MATLAB is essentially the same program whatever the operating system or computer, and this is one of its attractions. The programs (M-files) included in this text were written for version 4 (for WINDOWS 3.1) but have been tested on version 5, and modified where necessary.

MATLAB is basically a program designed to perform matrix arithmetic quickly and easily, and display the results of computations as high quality graphics. MATLAB is a very high level computer language, already programmed to carry out many complex activities, such as matrix multiplication, fitting curves to data, calculating Fourier transforms, and producing two- or three-dimensional plots of data, including contour maps.

MATLAB is an *interpreter*, i.e., each command in the program is interpreted and implemented, before the program goes on to the next command. Despite this, it is fast and easy to use to perform scientific calculations, because it is possible to write short MATLAB *scripts* and *functions*, which are the equivalent of long programs or subroutines in a language such as FORTRAN or Pascal, and because the MATLAB 'commands' are optimized for accuracy and speed. MATLAB also looks after all the problems of memory management, and most of the scaling problems involved in producing graphic output. In short, it enables the scientist to concentrate on obtaining the solution to a numerical problem, without being much concerned about the numerical methods used, or the details of input, output, and memory management. Taking care of these matters takes up about 90 percent of a scientist's time when he programs in a conventional, low-level computer language.

Most computer languages support many different data types, for example, booleans, strings, integers, and real numbers (but they often do not support vectors, matrices, and complex numbers). MATLAB basically supports only a single data type: the *matrix*. A matrix is an array of numbers: a single number (scalar) is a 1×1 matrix, a row or column of n numbers (vector) is a $1 \times n$ or $n \times 1$ matrix, so scalars and vectors are just special cases of matrices. Complex numbers have a real and imaginary part, so may be treated as a special type of element in a matrix. Matrices may be added, subtracted, multiplied, and manipulated algebraically or numerically using the rules of *matrix algebra*. MATLAB not only implements the *matrix operations* of ordinary matrix algebra, but adds a very useful set of *array operations*, which carry out operations element-by-element, rather than matrix by matrix.

Students who have not yet studied matrix algebra often believe that this fact will make it difficult for them to learn MATLAB. This is not the case: in learning MATLAB, a student will learn some linear (or matrix) algebra, which the student will certainly need anyway in order to become familiar with the techniques of data analysis, but it is *not* necessary to be familiar with matrix algebra in order to learn MATLAB.

A.2   Starting MATLAB

The procedures for starting MATLAB depend upon the computer you are using, and whether or not you are sharing it with other users. Read the Manual carefully. If you are using a PC, you must first enter WINDOWS, and start the program by double clicking on the MATLAB icon. This calls a file `matlabrc.m` which defines the default directories and the `path` that MATLAB follows to find files.

After MATLAB is installed from WINDOWS, you can see the current path by typing `path`, and you can add paths by following the procedure given in the Manual. If you prefer, use the MATLAB `help` command to get detailed information about any function or script (whether or not it forms part of the MATLAB standard set). Another useful function is `lookfor`: for example, typing `lookfor path` not only gives a brief description of the `path` function but also of many other functions that have `path` in their name (e.g., `addpath`). The current directory can be changed using the DOS `cd` command. If you are sharing the computer with others, you may want to use one of the diskette drives as the current directory, so that your data files and MATLAB scripts will be read from and written to your own diskette. If you do this, you must have a diskette in the designated drive. Alternatively, you may want to create your own subdirectory on the hard drive to store your files. The contents of the directory can be seen using the DOS `dir` command.

A.3   ASCII Files

ASCII is one of the few universally accepted standards in the computer industry. Letters and numbers are stored by computers as binary codes, but it would be very inconvenient if we had to type them in that form. Instead, we type them using a normal keyboard, and the computer converts the keyboard symbols into binary code, using the ASCII convention. A table showing the ASCII codes is available in almost every computer reference manual, and can be displayed on the screen by many editors. The first 128 symbols (0–127) are absolutely standard, the second 128 symbols vary somewhat from country to country.

The fact that ASCII is a universal standard means that ASCII files are a convenient way to prepare text or numbers that are to be read into a computer. A program that prepares files using only ASCII symbols is called an *editor*. Such files are often called *text files*, even though they may contain numbers and other symbols not present in ordinary text. A further advantage of such files is that they may be sent by EMail. Other types of files (called *binary files*) must be transmitted using special routines (e.g., `ftp`) or converted to text files by a program, such as `UUENCODE`, before they can be transmitted by EMail.

Many programs, including MATLAB, save data in formats other than ASCII. One reason for this is that less disk storage space is required. All programs are, however, capable of producing output in ASCII format. If disk storage is not a problem, then saving data in ASCII will avoid incompatibility problems in the future.

To use MATLAB, you will also need to use an editor. Many people already have a favorite editor, and if so, there is no reason to change it. If you are using DOS, the MS-DOS editor is easily invoked by typing `edit <filename>`. In WINDOWS, you can use Notepad. The WINDOWS Notepad can be invoked from within MATLAB to prepare new M-files, or edit existing ones. You can even use your favorite word processor provided you remember to save the file as a text file.

A.4 Entering Data

The methods for entering data are described in the MATLAB Tutorial. Data can be entered from the keyboard, loaded using `load`, or loaded by using an `M-file`.

- The following examples show how to enter data from within MATLAB. For a single number, e.g., 6.3, type

```
a = 6.3;
```

Typing the semicolon after the number stops MATLAB from immediately verifying the input by re-displaying it on the screen.

To enter a row vector, type

```
x = [6.3  7.2  -0.3  4.8];
```

To enter the same data as a column vector, either place a semicolon after each number, or convert the row vector to a column vector by typing

```
x = x';
```

This is an example of the *transpose* operation, which converts rows to columns of a matrix (and vice-versa).

To enter a matrix, type

```
X = [6.3  7.2;  -0.3  4.8; 7.9  -2.4];
```

The semicolons separate rows, so this is a 3×2 matrix. Note that it is assigned to **X**, not **x**. MATLAB is *case-sensitive*, that is, upper and lower case letters refer to different quantities. It is useful to form the habit of using capitals for matrices, and lower case letters for vectors or scalars—but it is not required in MATLAB. Nor is it necessary to declare, before using a variable, whether it is a matrix or a vector, or what size it is. Changing the size of a matrix takes time, however, so if a matrix is to be accumulated bit-by-bit by computation, it is best to declare its final size before computation begins. This is done by filling the matrix full of zero elements: for example, for a 20×4 matrix, use

```
X = zeros(20,4);
```

MATLAB provides many other shortcuts for creating other types of matrices—see the Manual for details.

- MATLAB's `load` command loads data from a file previously created and saved to disk. The file may be a `MAT` file, created by MATLAB (using the `save` command), or an ASCII file. Suppose it is an ASCII file saved as `c:\temp\mydata.dat`. If this directory is not the current directory type

```
load c:\temp\mydata.dat
```

MATLAB assigns the data to a variable called `mydata`— it can be displayed by simply typing that name. A file `hdrload.m`, available from The MathWorks, Inc. but not part of the regular MATLAB program, allows for loading ASCII files that have a header that identifies or describes the data. See the technical note at `www.mathworks.com/support/tech-notes/1400/1402.html`. To use `load`, however, the file must consist only of rows and columns of data, separated by spaces. The data may be in decimal form, or in exponential format. For example, either `0.0023`, or `2.3e-3`.

- Data may also be stored in `M-files`. These are ASCII files that designate the variable as well as the numbers. For example, `mydata.m` might contain

```
    % An example of data in an M-file
M = [1 4 7; 2 8 9]
```

Typing `mydata` loads and displays the data in this file, assigned to the matrix M. Note that this method has the advantage that a description of the data may be included as a comment—in MATLAB M-files, all lines (or parts of lines) preceded by a % are comments that are ignored when the instructions in the file are implemented.

Several other ways to enter data are described in the Manual, but the ones listed above are generally the most useful.

A.5 Matrix and Array Operations

The basic operations of matrix algebra are transposition, addition and subtraction, multiplication, and calculation of the determinant or inverse. The mathematical definitions are described in texts on linear algebra (e.g., Strang, 1980) and summarized in many other works (e.g., Davis, 1986). Their implementation in MATLAB is described and illustrated in Chapter 4 of this text, and in the tutorial provided in the MATLAB Manual. The brief remarks that follow point out some of the extensions to normal matrix operations implemented by MATLAB.

One set of extensions are MATLAB's *array operations*. Matrix addition and subtraction are carried out by adding or subtracting the corresponding elements of two matrices of the same size, i.e., with the same number of rows and columns. Matrix multiplication, however, is a complex operation (see Chapter 4) that involves much more than multiplying corresponding elements. Matrices may be raised to integral powers by repeated multiplication. Matrix division is not defined in ordinary matrix algebra.

MATLAB defines element-by-element multiplication, division, and exponentiation, carried out using the "dot operators"

```
C = A .* B;
C = A ./ B;
C = A .^p;
```

Use of these array operators is one way that one can avoid using slow-running program loops in MATLAB (see below).

MATLAB also defines two types of division. Backslash, or matrix left division, is one of MATLAB's most useful functions. Suppose we have a matrix equation **Ax = b**, which we wish to solve for **x**. This is done in MATLAB simply by typing

```
x = A\b;
```

If **b** is a column vector with the same number of rows r as the square matrix **A**, then **x** is a column vector giving the solution to the equation, determined by a fast numerical method that does *not* involve finding the inverse of **A**. **b** could be a matrix with r rows and c columns: in this case the solutions are found for each column of **b**. If **A** is not square, then there is no solution in ordinary matrix algebra, but MATLAB determines the least square solution (see Chapter 5) for the under- ($r < c$), or over-determined ($r > c$) set of equations. An array form of left division is also defined. See the Manual for details.

Slash, or right division, is rarely used: see the Manual for its definition.

MATLAB makes it easy to generate several special types of matrix. Examples are

- The null matrix, with all elements equal to zero, r rows, and c columns, is generated by X = zeros(r,c).

- The unit, or identity, matrix, a square matrix with zeros everywhere except for ones in the principal diagonal, is generated by I = eye(n).

- A matrix with ones in every element is generated by Y = ones(r,c).

- An index vector, with elements consisting of the integers from 1 to n is generated by
  `i = [1:n]`.

- A vector consisting of the pth column of the matrix **A** is generated by `u = A(:,p)`,
  and a vector consisting of the qth row by `v = A(q,:)`.

MATLAB also has several functions that are useful for determining the properties of matrices, and changing them. The following are examples:

- The size of a matrix, or length of a vector may be determined by the functions `size`,
  or `length`

- The organization into rows and columns may be changed using `reshape`. For convenience, a set of 100 measurements on a single variable may be prepared as a 20×5
  data table. MATLAB interprets this as a matrix (say A) with 20 rows and 5 columns,
  but for calculation purposes it must be converted into a column vector b. This is easy
  to do using

  ```
  b = reshape(A, 100, 1);
  ```

  If the size of A was not already known, we could use the following two lines instead

  ```
  [r,c] = size(A);
  b = reshape(A, r*c, 1);
  ```

- Suppose we want to rearrange the rows of a column vector u. There are five rows,
  which should be rearranged in the order given by the index vector `i = [4 2 3 5`
  `1]`. The result is obtained using `v = u(i)`. (It does not matter if u is a row rather
  than a column vector, the rearrangement is not affected.)

- MATLAB also provides a useful technique, known as *logical subscripting* to select
  only certain elements from a matrix. For example, the following commands select
  only the positive elements in a vector from a:

  ```
  b = a>0;
  c = a(b);
  ```

  If there are five elements in a and the second and fourth are negative, `b = [1 0 1`
  `0 1]`. Use of this vector in the second statement selects only those elements with
  $b(i) = 1$.

In a later section, we will see how these matrix functions can be used to "vectorize" MATLAB scripts and functions, thereby increasing the execution speed.

A.6 Scripts and Functions

The nature of MATLAB "programs," (generally called scripts or functions) is discussed in Chapter 3, and compared with programs written in languages such as C or Pascal. Both scripts and functions are sequences of MATLAB commands that are composed and saved as an M-file (e.g., myfile.m) before being invoked by typing the name of the file (e.g., myfile). There are some important differences: parameters cannot be passed to or from scripts, so all variables are global. Functions are compiled and stored in memory the first time they are called, so they do not need to be compiled again. So in general, it is better to write programs as functions than as scripts.

The fact that a function is compiled at the first call still does not increase the speed much, if the function has loops using the for or while commands: these will still be slow. If loops are unavoidable (e.g., as in logist in Chapter 10), then it is worth considering whether or not to compile the function as a MEX file. To do this, the function must be rewritten in C or FORTRAN, and compiled following the instructions in the MATLAB Manual.

A.7 Vectorizing Code

Changing MATLAB scripts or functions to replace loops with vector or matrix operations is called *vectorizing the code*. As this can often increase the speed of execution by more than an order of magnitude it is generally well-worth doing, but there are exceptions:

- In many cases, it is not at all obvious how to vectorize code. It is a skill that depends upon a thorough familiarity with both vector and matrix algebra, and MATLAB commands. If the existing code does not use the loop frequently, and runs fast enough for your purposes, why spend hours trying to vectorize it? Generally programming time is more valuable than machine time.

- Sometimes the meaning of vectorized code is obscure. For example, what does this code segment do?

```
v = [1:5]';
M = v(:,ones(3,1))
```

This is an example of a method for duplicating a vector, well-known to expert MATLAB programmers (but not to anyone else), and called "Tony's Trick" (see MATLAB's tech-note 1109 "How Do I Vectorize My Code?").

The answer is

```
M =
     1     1     1
     2     2     2
     3     3     3
     4     4     4
     5     5     5
```

Its meaning is much less obvious than

```
for j = 1:3
   M(:,j) = v;
end
```

which does the same thing. Sometimes it is a lot more important to write code that people (including the code's author, a few weeks later!) can understand, than code that saves a few milliseconds of execution time.

With these disclaimers in mind, we nevertheless present a few examples of vectorized code.

- Suppose we want to add the numbers in a column vector, **x**? If you know MATLAB, you know you can do this using `s = sum(x)`. But supposing this built-in function did not exist? In low level languages one would use a `for` loop, as illustrated in Chapter 3. MATLAB can do this too, but how can the operation be vectorized? Recall the definition of multiplying two vectors: if there are p rows in **x** then two products with a row vector **y** are possible, so long as **y** has p columns. One is **xy**, which produces a $p \times p$ matrix, the other is **yx** which produces a scalar. If **y** is a row of p ones, then **yx** is the sum of the x_i. So one answer to our vectorization problem is

```
y = ones(size(x))';
s = y*x
```

  Try this on your own computer using `vtest1.m`. On mine, MATLAB's `sum` was only a little faster than the vectorized version given above, and both were about 50 times faster than using a loop.

  Because the `sum` function exists, this example may seem a little academic—but the same logic may be applied to computing sums of cross-products (e.g., $\sum x_i y_i$).

- Suppose that you have a matrix **X** whose first two columns consist of x and y coordinates (other columns include measurements made at the location with those coordinates). You want to select only the data from a area whose coordinates lie between x_1, x_2, y_1, and y_2. The normal programming method would be to use a `for loop` to scan through the rows of the data, and an `if` statement to select the rows that satisfy the conditions set. This would be very slow in MATLAB, but can be avoided using the logical subscripting technique explained in the earlier section on Matrix and Array Operations.

```
m = (X(:,1)>=x1) & (X(:,1)<=x2) & (X(:,2)>=y1) &...
    (X(:,2))<=y2);
Y = X(m>0,:);
```

  `vtest2.m` compares the speed of this operation with one involving `for` and `if` statements. On my computer, the vectorized method was about 8 times faster.

- As a further example, compare the following vectorized version of Edist with the version given in Chapter 6.

```
function D2 = edist2(X)
% D2 = edist2(x)
% calculates the standardized Euclidean distance
% for the N samples in the rows of the data
% matrix X.  This version is vectorized.
% Written by Gerry Middleton, Feb. 1997
[N c] = size(X);
D = zeros(N);
D2 = zeros(N);
k = [1:N];
i = reshape(ones(N,1)*k,1,N^2);
j = reshape((ones(N,1)*k)',1,N^2);
D = sum((X(i,:)' - X(j,:)').^2)/c;
D2 = reshape(D2,N,N);
```

The problem here can be seen by examining the formula to calculate the Euclidean distance:

$$d_{ij}^2 = \left(\sum_k (x_{ik} - x_{jk})^2 \right) /n \tag{A.1}$$

This equation involves 3 indices, so to be converted to a matrix equation it really needs a three-dimensional matrix. We get around this by expanding the difference matrix by using expanded vectors of indices. For example, suppose there are 4 samples and 3 variables. Then

```
i = [1 1 1 1 2 2 2 2 3 3 3 3 4 4 4 4]
j = [1 2 3 4 1 2 3 4 1 2 3 4 1 2 3 4]
```

(Note the tricky way we generate these vectors in MATLAB.) Then calculating a set of differences using

```
D = X(i,:) - X(j,:)
```

yields a matrix with 16 rows and 3 columns. The first 4 rows give the differences between sample 1 and samples 1–4, the second four rows give the differences for sample 2, and so on. To calculate the distance, we need first to calculate the square of these differences, then take the transpose, so that we can use sum to find their (normalized) sum for each sample. This gives us a vector with 16 normalized sums, and we then reshape this to get the matrix we want. In other words, we accomodate our "three-dimensional" difference matrix, by making a two-dimensional matrix consisting of four (4×3) submatrices: after summing over the three variables, this reduces to a vector, which we then rearrange as a 4×4 matrix.

For all our efforts, this vectorized version seems to run slightly slower than the original version.

This last example is, perhaps, sufficient to illustrate that vectorizing MATLAB code is an art rather than a science, and does not always produce the desired results.

A.8   Figures

One of the advantages of MATLAB is that it produces high-quality figures with a minimum of effort by the programmer. It is inevitable, however, that, sooner-or-later, the user will want to modify the figures that MATLAB produces using the `plot` command. Simple matters are covered by several other MATLAB functions: titles are produced by `title`, axis labels by `xlabel` and `ylabel`, text can be added to a figure using `text`, several figures can be printed on the same page by `subplot`, logarithmic axes can be produced by `semilogx`, `semilogy` or `loglog`. The scaling of axes is controlled by `axis`. Grid lines may be turned on or off using `grid`. All of these functions are explained in the MATLAB Tutorial and Manual. Some of the less standard graphics displays, such as `bar`, `histo`, `contour`, and `mesh` are explained and illustrated as they are used in the book.

In what follows, we expand on some ways that the standard figures may be modified. Some of these techniques are only partly documented in the Tutorial and Manual: a more complete reference is given by Hanselman and Littlefield (1996, 1998). In the Manual, the most complete documentation is found under `axes`; see also the MATLAB online help.

MATLAB calls its system of low-level functions for controlling graphics Handle Graphics^{TM}, and gives a brief discussion in the Tutorial. Figures and axes are examples of *objects*, which are identified using a number called a *handle*. There is a heirarchy of objects. For example, figures include axes, and within axes, the font name and size, and the line width are objects with their own handles. Handles are automatically assigned when an object is created, together with their default properties (called *factory settings* in MATLAB). You can determine what the handle of the current figure or axes is by typing `gcf` or `gca`. Typing `get(gca)` shows not only the current axes handle, but all the objects in the axes class, together with their settings.

While creating a figure, you can assign the handle to a variable, h, by the command `h = figure;`. (`figure` is a useful command for another reason—if your function makes more than one figure, and you wish to examine (and possibly print) each one after the function has been run, use `figure` before each graphic is created, and then each graphic will be produced in a separate figure window.)

The following are examples of how you can control some aspects of figures using Handle Graphics:

- MATLAB supports several different fonts which can be used in graphics. The default font is `Helvetica`. Though MATLAB supports eleven fonts, your computer may not. Try running the following demo:

```
% script fontdemo.m
h = text(0.1,0.9,'This is the default font: Helvetica');
axis off;
hold on;
h1 = text(0.1,0.8,'This is Courier font text');
set(h1,'FontName', 'Courier');
h2 = text(0.1,0.7,'This is Times font text');
set(h2,'FontName', 'Times');
h3 = text(0.1,0.6,'This is Avante Garde font text');
set(h3,'FontName','Avante Garde');
h4 = text(0.1,0.5,'This is Bookman font text');
set(h4,'FontName','Bookman');
h5 = text(0.1,0.4,'This is Helvetica Narrow font text');
set(h5,'FontName','Helvetica Narrow');
h6 = text(0.1,0.3,'This is New Century Schoolbook font text');
set(h6,'FontName','New Century Schoolbook');
h7 = text(0.1,0.2,'This is Palatino font text');
set(h7,'FontName','Palatino');
h8 = text(0.1,0.1,'This is Zapf Chancery font text');
set(h8,'FontName','Zapf Chancery');
```

Note that text places the text, beginning at a location specified by the x and y coordinates: if the coordinates of the figure have not already been designated (e.g., by a plot command) then the default x and y range from 0 to 1.0.

The following example shows how to control the font used in titles or axis labels:

```
title('The title of the graph');
h = get(gca,'title');
set(h, 'FontName', 'Times');
```

Note that the axes of the figure have one handle (referenced by gca) but the title is an object in the axes class that has its own handle. The example shows how to retrieve this handle, so that the font in the title can be changed.

Another way to get access to different fonts is to use The Styled Text Toolbox, written by D.M. Schwarz, which may be downloaded from The MathWorks, Inc. ftp site (search for stextfun. This package permits easy composition of mathematical or chemical formulas (using a TEX -like language), for display on MATLAB figures. Without this toolbox it is very difficult to include subscripts in MATLAB figures.

- For scientific applications, Greek letters are probably the most important ones that are not available in regular fonts. They are supplied by the symbol font, and can be identified using the following demo:

```
% script fontdem2
% displays roman and greek fonts
text(0.1, 0.8, ['a b c d e f g h i j k l m n o p q '...
 'r s t u v w x y z']);
hold on;
axis off;
h = text(0.1, 0.7,['a b c d e f g h i j k l m n o p q '...
 'r s t u v w x y z']);
set(h,'FontName','Symbol');
text(0.1, 0.6,['A B C D E F G H I J K L M N O P Q '...
 'R S T U V W X Y Z']);
h = text(0.1, 0.5,['A B C D E F G H I J K L M N O P Q '...
 'R S T U V W X Y Z']);
set(h,'FontName','Symbol');
hold off
```

For example, to label the x and y axes with the Greek letters σ and τ, use xlabel('s'),
ylabel('t'), followed by hx = get(gca, 'xlabel'), hy = get(gca, 'ylabel')
and set(hx,'FontName','Symbol'), set(hy, 'FontName','Symbol').

- Line widths and symbol sizes can be controlled on graphs, as shown by the following
  example (the scale is in *points*, where a point is about 1/72 in):

```
% script sizedemo.m
x = [1 2 3 4 5];
y = [1.1 1.9 2.8 4.1 4.9];
subplot(221);
plot(x,y,'o',x,y);
subplot(222);
h1 = plot(x,y,'o',x,y);
set(h1,'MarkerSize',3,'LineWidth',0.5);
subplot(223);
h2 = plot(x,y,'o',x,y);
set(h2,'MarkerSize',7,'Linewidth',1);
  subplot(224);
h3 = plot(x,y,'o',x,y);
set(h3,'MarkerSize',10,'LineWidth',1.5);
```

The result of running this demo is shown in Figure A.1. (The appearance on the
screen may be quite different from the printed graph.) Note that the Linewidth
setting affects not only the plotted lines, but also the width of the lines used to form
the symbols. To produce different sizes of marker, without changing the line width
used to draw them, use a separate plot command.

- The ticks and labels on axes can be customized. For an example of how this is done
  see the listing of cumprob in Chapter 2.

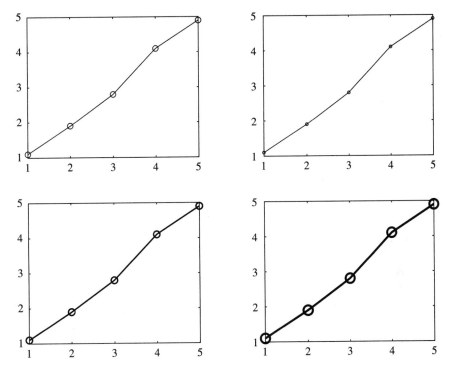

Figure A.1   Four subplots produced using different values for MarkerSize and LineWidth. Top left uses default values; top right uses 3 and 0.5; bottom left uses 7 and 1; bottom right uses 10 and 1.5.

- The size of graphs may be customized: this is particularly useful if you wish to have original printouts for publication, or for pasting into a document. The technique is illustrated in the following script:

```
% script mplot3.m
% sets up a figure so that it is
% 3in high, and printed 2in from left margin
% and 3in above bottom of paper
% fonts are scaled to 10 points
set(gcf, 'PaperUnits','inches','PaperPosition',[2 3 5 3]);
set(gcf, 'Units', 'inches','position',[1.1 1.1 4.8 2.8]);
set(gcf, 'DefaultAxesFontSize',10);
x = 0:0.1:6*pi;
plot(x,sin(x),':', x, cos(x));
title('Sine and Cosine');
xlabel('radians'), ylabel('amplitude');
legend('sine    ','cosine    ',-1);
```

The object `PaperPosition` specifies the left margin, lower margin, width, and height of the figure (in that order). The object `position` specifies the actual position of the graph: some extra space must generally be left for title and axis labels. `legend` is a feature not well documented in some Manuals (though it is in the Student Edition). See the online `help` for details. It labels the curves plotted: the final argument specifies the "tolerance"—if this is 0 then it will be plotted inside the plot only if the legend box does not cover any points; if it is -1, then the legend box is forced outside the plot. The result is shown in Figure A.2.

Note that what is displayed on the screen is not always exactly what you will get when you print the graphics: in particular, the appropriate line widths to use should probably be determined after a few trial runs with your printer.

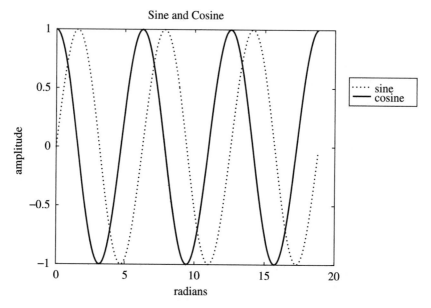

Figure A.2   Sine and Cosine curves, with a legend

MATLAB figures may be saved as a Postscript file for inclusion in other documents, using the `print` command. A variety of formats are available: see the Manual for details. The appropriate one to use depends on the software producing the document. MATLAB does not provide a graphics editor, but one (called `MatDraw`) is available as Shareware from The MathWorks, Inc. `ftp site`.

A.9 Recommended Reading

Much useful information about MATLAB (and its various clones) can be found in the MATLAB FAQ. This is available from the MATLAB `ftp` site `ftp.mathworks.com`, or their Web site `http://www.mathworks.com`.

Hanselman, D., and B. Littlefield, 1996, Mastering MATLAB. Englewood Cliffs NJ, Prentice-Hall, 542 p. (QA297.H298 By the authors who prepared the Student Manual— a comprehensive guide.)

Hanselman, D., and B. Littlefield, 1998, Mastering MATLAB 5. Englewood Cliffs NJ, Prentice-Hall, 638 p. (QA297.H296 For users of version 5.)

Strang, G., 1980, Linear Algebra and Its Applications. Englewood Cliffs NJ, Prentice-Hall, second edition. (A clear introduction to matrix algebra.)

Appendix B

List of Scripts, Functions, and Data Files

B.1   Scripts and Functions

The following pages list the scripts and functions described in this book, and the page reference. All of these (except `griddat` and `triangul`) are included in the diskette that accompanies the book: most, but not all, are listed in the book. Regular MATLAB functions are not listed.

arc	166	diffuse2	197
arc2	167	discpl2	86
barth	186	discplot	89
betafn	76	divider2	206
boxcnt	207	divplot	201
centax2	163	divplot2	202
contvec	136	doublev	96
costheta	85	edist	84
cotomat	146	edist2	243
cump	21	embed	218
cumpinv	21	fdiff	103
cumprob	27	flrnz	213
dattogr	146	fontdemo	245
dendro5	96	fontdemo2	246
dendrox2	96	fsprb	216
deseas	123	gridcon	143
diagzero	96	griddat	141

B.2   Data Files

The following pages list the files containing data in `ascii` format. M-files can be loaded directly by typing the file name; `.dat` files must be loaded using the function `load`. Where no reference is given, the data was generated by the author.

File Name	page	Reference
apxyz.dat	136	Isaacs and Srivastava, 1989
ar.m	188	
asmax2.dat	56	Krumbein, 1942
asmean.dat	56	Krumbein, 1942
bedf.dat	78	
brooker.dat	151	Brooker, 1991
brown.dat	207	
cvcrk.dat	106	Haan, 1977
dewijs.dat	5	De Wijs, 1951
dgrid.dat	134	
dmap.dat	138	Davis, 1986
dstcl.dat	75	Draper and Smith, 1981
ffeld.dat	163	Fisher, 1993
fleb2.dat	171	Fisher at al., 1987
fleb4.dat	171	Fisher at al., 1987
gal.dat	63	Shaw and Bankier, 1954
iriset.dat	88	Fisher, 1936
irisvers.dat	88	Fisher, 1936
iskrig.dat	157	Isaacs and Srivastava, 1989
iskrig2.dat	157	Isaacs and Srivastava, 1989
koch5.dat	205	
lit1.dat	33	Moss et al., 1995
lit2.dat	33	Moss et al., 1995
lit3.dat	52	Moss et al., 1995
lrnz50.dat	218	
mauna.dat	103	Garcia, 1994
mibas.dat	70	Clayton et al., 1966
molr.m	187	
nrex.dat	188	Russell and Stanley, 1990
os.m	187	
plates.dat	68	Forsyth and Uyeda, 1975
prdav.dat	107	Savini and Bodhaine, 1971
purdyr.dat	98	Purdy, 1963

Index

Items in `typwriter font` are MATLAB functions, mostly written by The MathWorks, Inc. Functions written for this book are listed separately in Appendix B, and data files are listed in Appendix C.

SYSTEM REQUIREMENTS.

The enclosed disk contains data which can run on any Windows® computer equipped with MATLAB 4, 5, or Student Version. At a minimum, a PC 386 with Windows® 3.1 (MATLAB 5 requires Windows®.95)